Chemistry Research and Applications

Chemistry Research and Applications

Pyrimidines and their Importance
Roger G. Ward (Editor)
2023. ISBN: 979-8-88697-656-4 (Softcover)
2023. ISBN: 979-8-88697-663-2 (eBook)

What to Know about Lanthanum
Catherine C. Bradley (Editor)
2023. ISBN: 979-8-88697-615-1 (Softcover)
2023. ISBN: 979-8-88697-623-6 (eBook)

The Essential Guide to Alkaloids
Deepak Kumar Semwal, PhD (Editor)
2023. ISBN: 979-8-88697-456-0 (Hardcover)
2023. ISBN: 979-8-88697-526-0 (eBook)

The Future of Biorefineries
Waldemar Nyström (Editor)
2023. ISBN: 979-8-88697-524-6 (Hardcover)
2023. ISBN: 979-8-88697-528-4 (eBook)

Introduction to Multidisciplinary Science in an Artificial-Intelligence Age: The Matter in our Universe, Biological Cells, and Plate Tectonics
Luc Ikelle
2023. ISBN: 978-1-68507-992-5 (Hardcover)
2023. ISBN: 979-8-88697-498-0 (eBook)

More information about this series can be found at
https://novapublishers.com/product-category/series/chemistry-research-and-applications/

Jeff C. Murdoch
Editor

The Chemistry of Gallic Acid and Its Role in Health and Disease

Copyright © 2023 by Nova Science Publishers, Inc.

All rights reserved. No part of this book may be reproduced, stored in a retrieval system or transmitted in any form or by any means: electronic, electrostatic, magnetic, tape, mechanical photocopying, recording or otherwise without the written permission of the Publisher.

We have partnered with Copyright Clearance Center to make it easy for you to obtain permissions to reuse content from this publication. Please visit copyright.com and search by Title, ISBN, or ISSN.

For further questions about using the service on copyright.com, please contact:

Phone: +1-(978) 750-8400	Copyright Clearance Center Fax: +1-(978) 750-4470	E-mail: info@copyright.com

NOTICE TO THE READER

The Publisher has taken reasonable care in the preparation of this book but makes no expressed or implied warranty of any kind and assumes no responsibility for any errors or omissions. No liability is assumed for incidental or consequential damages in connection with or arising out of information contained in this book. The Publisher shall not be liable for any special, consequential, or exemplary damages resulting, in whole or in part, from the readers' use of, or reliance upon, this material. Any parts of this book based on government reports are so indicated and copyright is claimed for those parts to the extent applicable to compilations of such works.

Independent verification should be sought for any data, advice or recommendations contained in this book. In addition, no responsibility is assumed by the Publisher for any injury and/or damage to persons or property arising from any methods, products, instructions, ideas or otherwise contained in this publication.

This publication is designed to provide accurate and authoritative information with regards to the subject matter covered herein. It is sold with the clear understanding that the Publisher is not engaged in rendering legal or any other professional services. If legal or any other expert assistance is required, the services of a competent person should be sought. FROM A DECLARATION OF PARTICIPANTS JOINTLY ADOPTED BY A COMMITTEE OF THE AMERICAN BAR ASSOCIATION AND A COMMITTEE OF PUBLISHERS.

Library of Congress Cataloging-in-Publication Data

ISBN: 979-8-88697-672-4

Published by Nova Science Publishers, Inc. † New York

Contents

Preface		vii
Chapter 1	**Gallic Acid: From Chemistry to Analysis**	1
	Jay Rana, Pooja Desai and Sonal Desai	
Chapter 2	**A New Perspective on the Efficacy of Gallic Acid in the Treatment of Breast Cancer**	31
	Sakshi Patil, Swapnali Patil, Pranali Pangam, Poournima Sankpal and Sachinkumar Patil	
Chapter 3	**A New Perspective on the Efficacy of Gallic Acid in the Treatment of Lung Cancer**	45
	Swapnali Patil, Pranali Pangam, Shravan Joshi, Poournima Sankpal and Sachinkumar Patil	
Chapter 4	**A New Perspective on the Efficacy of Gallic Acid Nanoformulation on Colorectal Cancer Treatment**	61
	Venketesh Kumbhar, Suraj Tarihalkar, Poournima Sankpal and Sachinkumar Patil	
Chapter 5	**Strategic Approaches of Gallic Acid in Different Disease Conditions**	77
	Komal Patekar, Sakshi Patil, Poournima Sankpal and Sachinkumar Patil	
Chapter 6	**Gallic Acid: A Potential Antidiabetic Agent**	93
	Suraj Tarihalkar, Venkatesh Kumbhar, Poornima Sankpal and Sachinkumar Patil	

Chapter 7	**Gallic Acid: A Potential Anti-Tumor Agent** ....109 Pranali Pangam, Swapnali Patil, Poournima Sankpal and Sachinkumar Patil	
Chapter 8	**In Vitro Anticancer Activity Gallic Acid Nanoparticles on Colon Cancer Cell Colo 205** ....129 Poournima Sankpal, Sachinkumar Patil, Pramod B. Patil, Rajanikant Ghotane, Prafulla Choudhari and Sanket Rathod	
Chapter 9	**Pharmacognosy of Gallic Acid and Its Co-Crystals** ....147 Sanchay Jyoti Bora, Riju Kakati Sarma and Purabi Sarmah	
Index	175	

Preface

This book explores the latest developments in gallic acid research and its role in health and disease.

Chapter 1 - Gallic acid, a polyphenolic acid, found in many medicinal plants, is a strong antioxidant compound. Gallic acid is reported to possess many biological activities such as antibacterial, anti-inflammatory, anti-cancer, etc. This simple and low molecular weight phenolic compound is analysed by various techniques namely Spectroscopy, Chromatography and Electrophoresis. The chemistry of the gallic acid plays a major role in its analysis. This book chapter highlights the chemistry of gallic acid with special attention on various analytical techniques adopted for its quantitative determination.

Chapter 2 - Gallic acid is (GA) a natural phenolic compound. It is also known as 3,4,5-trihydroxybenzoic acid. It has been suggested that the natural phenolic chemical gallic acid (3,4,5-trihydroxybenzoic acid; GA), which is produced from plants, can stop the growth and spread of a variety of malignancies. Anywhere in the body, cancer is the uncontrolled development of aberrant cells. Malignant, tumor, or cancer cells are all terms used to describe these aberrant cells.

The second biggest cause of mortality in the world is cancer. With more than a million new instances of breast cancer diagnosed worldwide annually, breast cancer is the most prevalent cause of death. Breast cancer is the most common cancer in women that accounts for 33% of all cancers in women globally. In 2020, 685000 people worldwide died and 2.3 million women were diagnosed with breast cancer. The most common disease in the globe as of the end of 2020 was breast cancer, which had been diagnosed in 7.8 million women in the previous five years.

For breast cancer and melanoma, there are several treatments available, including surgical removal of tumor, radiation therapy, hormone therapy, chemotherapy, targeted biological therapies, etc. Many of them have negative side effects. In order to meet the constant high demand for new anticancer medications, scientists investigate numerous natural and synthetic substances. Gallic acid (GA) found in many dietary substances and herbs used in traditional medicine. It has antibacterial, antiviral, anti-inflammatory, and antioxidant effects. A naturally occurring phenolic compound found in plants called gallic acid (GA) has a number of therapeutic effects that include anti-inflammatory, anti-obesity, and anti-cancer activities.

In more recent studies, gallic acid (GA) has been demonstrated to perform anti-cancer actions through a number of biological mechanisms, including angiogenesis, cell cycle arrest, migration, metastasis, apoptosis, and oncogene expression. In-depth research on gallic acid (GA), anticancer effects in MCF-7 human breast carcinoma cells was done for this review. This article indicated that treatment with gallic acid (GA) resulted in inhibition of proliferation and induction of apoptosis in MCF-7 cells. In conclusion, the authors' review shows that gallic acid (GA) is a unique, powerful, and safe anti-cancer therapeutic candidate for the treatment of breast cancer.

Chapter 3 - According to estimates from the International Agency for Research on Cancer, 1 in 5 individuals worldwide will acquire cancer at some point in their lifetime. As per United States study reports, it is estimated that there will be 1,918,030 new cancer cases and 609,360 cancer deaths in 2022, with lung cancer being the primary cause of death accounting for around 350 of those fatalities daily. Clinical interventions have had very little chance of reducing lung cancer-related deaths in recent years.

More than 80% of occurrences of lung cancer are non-small-cell lung cancer (NSCLC), which is the most prominent subtype, whereas small cell lung cancer (SCLC) makes up more than 15% of all occurrences of lung cancer and is the most dangerous subtype of the disease. A variety of molecular mechanisms and biological pathways play an important role in the development of lung cancer therapies. In order to prevent and treat non-small-cell lung cancer (NSCLC) and small cell lung cancer (SCLC), polyphenols are the naturally occurring sources of potential cancer chemotherapeutic agents that can minimize the side effects of conventional anticancer medications and aid in the fight against drug resistance.

Gallic acid (GA) is gaining significant interest as a form of phenolic acid due to its excellent bioavailability and non-toxicity. The most recent research on gallic acid's anti-tumor properties in various malignancies was examined, with an emphasis on the molecular mechanisms and cellular pathways that lead to tumor cell apoptosis and migration. When gallic acid and chemotherapeutic drugs are administered simultaneously, tumor proliferation is suppressed more effectively. In this chapter, the authors extensively studied the anticancer effect of gallic acid in human SCLC H446 cell line and NSCLC EGFR-TKI (tyrosine kinase inhibitor)-resistant cell line. This article indicated that treatment with gallic acid (GA) resulted in inhibition of proliferation and induction of apoptosis in NSCLC and SCLC cells. According to this investigation, evidence suggests that gallic acid is a promising new, potent, and safe anti-cancer therapeutic candidate for the treatment of lung cancer.

Chapter 4 - Cancer is a broad collection of illnesses that can begin in practically any organ or tissue of the body. These illnesses are brought on when abnormal cells grow out of control, cross their normal boundaries to infect nearby body parts and/or spread to other organs. An estimated 10 million deaths were attributed to cancer in 2020, making it the second highest cause of death worldwide. Men are more likely to develop lung, prostate, colorectal, stomach, and liver cancer than women, who are more likely to develop breast, colorectal, lung, cervical, and thyroid cancer.

A polyphenolic substance called gallic acid (GA) has been shown to prevent a number of disorders. In addition to prescription medications, nutraceuticals and medicinal foods are being used to treat neurological disorders, cancer, hepatitis C, inflammation, and cardiovascular diseases, even though the majority of these conditions have only been investigated *in vitro*. Colorectal cancer, which is the third most common disease worldwide and one of the main reasons why people die from cancer.

This chapter is based on research into various gallic acid nanoformulations, which illustrates their treatment resistance for colorectal cancer. The application of nanotechnology in medicine has had a significant impact on the development of theranostic agents, which may simultaneously diagnose and treat illnesses. Several nanocarriers have been created and successfully employed to carry pharmaceutical drugs, including graphene oxide, polymers-based delivery systems, layered double hydroxides, gold nanoparticles, multifunctional nanoparticles and iron oxide magnetite nanoparticles.

Chapter 5 - Gallic acid (GA) is a natural phenolic compound. It is found in plants and foods that have been esterified with sugars, also known as Gallo tannins. It is present in fruits, such as black currant, berries, grapes, avla, clove. It is said to have a number of health enhancing properties. It appears in plants as hydrolysable tannins, free acids, esters, catechin derivatives, and free acids. In numerous conditions where oxidative stress has been linked, such as cardiovascular diseases, cancer, neurological disorders, and ageing, it has been shown to have potential therapeutic and preventive effects.

In order to portray the pharmacological status of this compound for future studies, this chapter aims to summarize the pharmacological and biological activities of (GA), including anticancer, antioxidant, and anti-inflammatory, antidiabetic, anti-myocardial, anti-obesity, antimicrobial properties among this numerous data *in vitro* and animal models. The outcomes of this review may highlight the potential of this compound as a novel therapy approach for the conditions listed, either on its own or in combination with other medications or their analogues to enhance their effects.

Results of the present study demonstrate that GA and effective supplement, as an adjuvant therapy may be a very promising compound reducing mortality rate of diseases.

Chapter 6 - Diabetes mellitus (DM), which increases morbidity and mortality, has emerged as a global health issue. Gallic acid is a phenolic molecule with antidiabetic properties. The creation and activation of oxidative stress were originally linked to the destructive course of diabetes mellitus (DM). Inflammation and oxidative stress have both been proven to have a significant role in the pathological development of diabetes mellitus and its associated consequences.

Gallic acid (GA) was studied for its potential to treat diabetes mellitus using natural antioxidants. Gallic acid has consistently shown strong anti-inflammatory and antioxidative effects on metabolic illnesses. Gallic acid is commonly available in herbal forms and edible plants. Gallic acid is a potent compound which shows anti-diabetic effects. A summarization of derivatives is required.

The objective of this chapter is to highlight the latest theories and findings in the fields of oxidative stress and diabetes mellitus. For diabetes mellitus and its consequences, gallic acid functions as an anti-glycemic drug.

Chapter 7 - Tumors are the second biggest global cause of mortality. The latest research in the area is concerned with how cancer therapy resistance arises and how to combat or prevent it. According to the present predicament, innovative anti-cancer drugs are urgently required for treatment of cancer cells that are resistant to chemotherapy. Phytochemical's pharmacological properties and ability to target a variety of biological pathways play a crucial part in the development of cancer therapies.

For the prevention and treatment of tumors, natural phenolic substances such as gallic acid (3,4,5 trihydroxybenzoic acid; GA) gained popularity. Gallic acid (GA) is a polyhydroxy phenolic molecule, typically found in natural sources including amla, berries, grapes, apple peels, green tea, gallnut, sumac, tea leaves, wine, chestnut, and oak bark. The most recent research on gallic acid's anti-tumor properties in various malignancies was examined, with an emphasis on the molecular mechanisms and cellular pathways that lead to tumor cell apoptosis and migration. When gallic acid and chemotherapeutic drugs are administered simultaneously, tumor proliferation is suppressed more effectively.

Chapter 8 - The main objective of this study was to evaluate cytotoxic activity of gallic acid-loaded sodium alginate nanoparticles using GMO and poloxamer 407 toward human colorectal cancer cell lines as COLO 205. Gallic acid was successfully encapsulated into nanoparticles after being characterized with particle size entrapment effectiveness, FT-IR, X-ray diffraction, DSC, and loading content. The dialysis membrane approach was employed during *in vitro* drug release study at different pH levels (1.2, 4.5, 7.5, and 7.0) to mimic the GIT condition. After 24 hours 79.06% of drug release was accomplished in a sustained manner.

Results from the MTT assay on the human COLO 205 cell line showed that gallic acid nanoparticles had more significant anti-colon cancer activity, with an IC50 value of 6.99ug/ml. However, currently, no specific study on anti-cancer activity has been published on gallic acid nanoparticles by *in vitro* COLO 205 cell line model.

Chapter 9 - Gallic acid (GA), a naturally occurring low molecular weight polyphenolic compound, is known to possess tremendous health benefits. Its presence as a phytochemical in numerous plant species is of great biomedical significance. Studies have established the presence of GA in the fruit and fruit pods of *Caesalpinia coriaria*, *C. spinosa*, *Caesalpinia sappan*, *C. brevifolia* (Fabaceae); leaves of *Rhus typhina*, *R. coriaria* and leaf galls of *Rhus semialata* (Anacardiaceae); wood and bark of *Quercus* sp and *Castanea* sp (Fagaceae), wood galls of *Quercus infectoria* (Fagaceae); fruit of *Terminalia chebula* (Combretaceae) and leaves of *Ximenia americana* (Olacaceae) to name a few.

The fungal species *Aspergillus fischerii* MTCC 150 has been used to yield GA by the hydrolysis of tannic acid. Also, the bacterial species *Klebsiella pneumoniae* and *Corynebacterium* sp have been reported to produce it from the crude extract of tara gallotanin. This biologically relevant molecule (GA) has been found to possess various pharmacological activities like antioxidant, antibacterial, antitumor, antiviral properties etc. Besides these, GA has no toxicity or side effects even in large doses. Consequently, it has received much attention in the field of pharmacological research. This review presents a report on the natural occurrence of GA and its therapeutic properties on the basis of available literature.

Due to the presence of potential hydrogen bond donors and acceptor sites, GA appears to be an excellent candidate for co-crystal formation. On the other hand, low thermal stability, large particle size and poor solubility during absorption are some of the issues due to which pharmacological activities of GA are significantly diminished. The therapeutic effectiveness of these materials greatly depends on their solubility because poor solubility can cause low bioavailability of the same. Literature reports reveal that the solubility and dissolution rate of GA can be greatly enhanced by constructing GA-based co-crystals with the help of several co-crystal formers (CCFs). In recent times, an increasingly larger section of chemists have started looking into the structures of binary crystalline solids with a view of examining the enhanced pharmacological properties in them. A consolidated account of reports available in literature related to the co-crystals comprised of GA, their formation, structure and pharmacological activities has also been provided in this chapter.

Chapter 1

Gallic Acid: From Chemistry to Analysis

Jay Rana
Pooja Desai
and Sonal Desai
Department of Quality Assurance,
SSR College of Pharmacy, Silvassa,
Union Territory of Dadra & Nagar Haveli, Daman Diu, India

Abstract

Gallic acid, a polyphenolic acid, found in many medicinal plants, is a strong antioxidant compound. Gallic acid is reported to possess many biological activities such as anti-bacterial, anti-inflammatory, anti-cancer, etc. This simple and low molecular weight phenolic compound is analysed by various techniques namely Spectroscopy, Chromatography and Electrophoresis. The chemistry of the gallic acid plays a major role in its analysis. This chapter highlights the chemistry of gallic acid with special attention on various analytical techniques adopted for its quantitative determination.

Keywords: gallic acid, HPLC, HPTLC, NMR, UV spectroscopy

Abbreviations

^{13}C NMR	Carbon Nuclear Magnetic Resonance
^{1}H NMR	Proton Nuclear Magnetic Resonance
AR	Aromatic
C_{18}	Octadecyl Silane

In: The Chemistry of Gallic Acid and Its Role in Health and Disease
Editor: Jeff C. Murdoch
ISBN: 979-8-88697-672-4
© 2023 Nova Science Publishers, Inc.

C_8	Octyl Silane
CH_3COOH	Acetic Acid
cm	Centimetre
cm^{-1}	Reciprocal Centimeter
COOH	Carboxylic Acid
FTIR	Fourier Transform Infrared Spectroscopy
GC	Gas Chromatography
H^+	Hydrogen Ion
H_3PO_4	Orthophosphoric Acid
HPLC	High Performance Liquid Chromatography
HPTLC	High Performance Thin Layer Chromatography
IR	Infrared Spectroscopy
LC	Liquid Chromatography
LOD	Limit of Detection
log P	Partition Coefficient
log ε	logarithm Epsilon
LOQ	Limit of Quantitation
mg	Milligram
min	Minute
ml	Millilitre
mm	Millimetre
MS	Mass Spectrometry
ng	Nanogram
nm	Nanometer
NMR	Nuclear Magnetic Resonance
ODS	Octadecyl Silane
OH	Hydroxyl Group
OPA	Orthophosphoric Acid
pKa	Acid Dissociation Constant
PPM	Parts per Million
R_f	Retardation Factor
TFA	Trifluoroacetic Acid
TLC	Thin Layer Chromatography
UV	Ultraviolet Spectroscopy
v/v	Volume by Volume
λ_{max}	Absorption Maxima
μg/ml	Microgram per Millilitre
μl	Microlitre
μm	Micrometre

Introduction

Gallic acid, a trihydroxy derivative of benzoic acid (Manish Pal et al., 2018), exists as a free molecule as well as an element of tannins, viz., gallotannin (Fitzpatrick and Woldemariam, 2017). Gallic acid was first identified in plants by Carl Wilhelm Scheele, a Swedish German pharmaceutical scientist in 1786 (Fischer, 1914). Gallic acid and its derivatives are abundant in many plants such as blueberry, blackberry, strawberry, plums, grapes, mango, cashewnut, hazelnut, walnut, tea, wine, etc (Daglia et al., 2014). Gallic acid is chiefly derived from the shikimic acid pathway (Herrmann and Weaver, 1999). Figure 1 gives a glance of shikimic acid pathway for biosynthesis of gallic acid.

Various studies demonstrated that gallic acid possesses various pharmacological activities namely anti-inflammatory (Rosas et al., 2019), anti-fungal, anti-viral and anti-cancer (Fiuza et al., 2004; Sameermahmood et al., 2010). Gallic acid, a phenolic compound with potential antioxidant properties, is also used as standard for estimation of total phenolic content present in plant materials and extracts (Aruoma et al., 1993). It is readily soluble in most of the polar solvents making it suitable to be analysed by many conventional as well as sophisticated analytical techniques. This chapter covers in depth review of gallic acid with reference to its chemistry and analysis.

Chemistry and Physico-Chemical Properties of Gallic Acid

Gallic acid is a crystalline solid, white or pale yellow in colour with a chemical name 3, 4, 5-trihydroxybenzoic acid (Choubey et al., 2018) and having molecular formula of $C_7H_6O_5$. The weight of one mole of gallic acid is 170.11954 grams, it melts at 250°C (Goldberg and Rokem, 2009) and has density of 1.694 g/cm^3 (Fernandes and Salgado, 2016). Gallic acid has four potential acidic protons and because of that gallic acid possess pKa values of 4.0 (due to COOH group), 8.7, 11.4, and >13 (due to phenolic OH) (Abbasi et al., 2011, Eslami et al., 2010). It gets solubilized in water, alcohol, ether and glycerol but is practically insoluble in benzene, chloroform and petroleum ether (Fernandes and Salgado, 2016). A study revealed that the solubility of gallic acid is highest in methanol, followed by ethanol, water and ethyl acetate (Daneshfar et al., 2008). Partition coefficient (log P) value for gallic acid is 0.70.

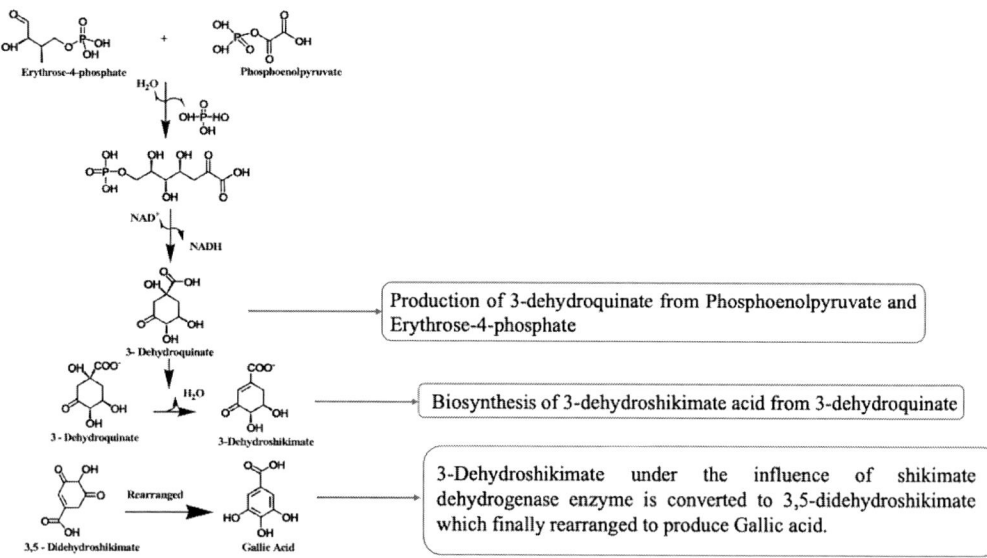

Figure 1. Shikimic acid pathway for biosynthesis of gallic acid.

Analytical Perspective of Gallic Acid

Ultraviolet Spectroscopy

UV-visible spectroscopy which uses UV and visible light in the wavelength ranging from 200 nm to 400 nm (Gandhimathi et al), is a widely utilized technique for the quantification of chemical compounds based on 'Beer-Lambert law' (Luis Aleixandre-Tudó et al., 2017). Gallic acid absorbs UV light and is suitable to be quantified by UV spectrophotometry (Aleixandre et al., 2013) due to the presence of phenolic rings (Boulet et al., 2017). Gallic acid exhibits absorption maxima at 272.5 nm with log ε 4.06 (Lide et al.). The absorption maxima are an important parameter for quantification by this technique which can be predicted based on molecular structure by Woodward-Fieser Rules by correlating the position and degree of substitution of chromophore. For gallic acid, base value for aromatic acid is 230 nm, presence of OH at ortho and meta position adds 14 nm (2 x 7) and OH group at para position adds 25 nm resulting in calculated absorption maxima of 269 nm. Figure 2 illustrates calculation of absorption maxima of gallic acid by Woodward-Fieser Rules.

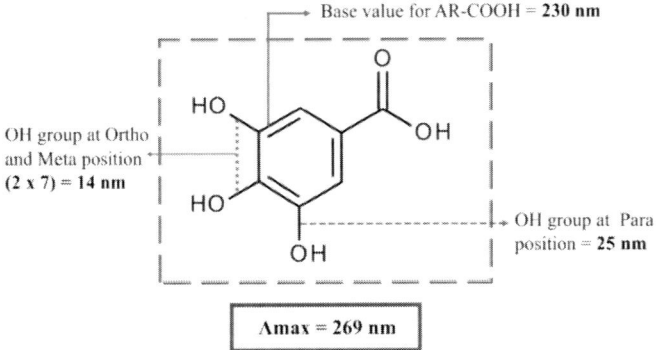

Figure 2. Calculation of absorption maxima of gallic acid by Woodward-Fieser Rules.

Many reported literatures are available for estimation of gallic acid by UV spectrometry. A summary of reports that have been published on the analysis of gallic acid by UV spectrometry is presented in Table 1.

Solvent used for UV analysis should be UV transparent. Solvents which do not contain conjugation are preferred for this technique. Methanol and water are commonly used as a solvent for estimation of gallic acid because of their low UV cut-off 205 nm and 190 nm, respectively. Bathochromic and hypsochromic shifts can be observed based on solvents used. Water is relatively more polar than methanol which causes the shifting of electronic transition in gallic acid to shorter wavelength leading to bathochromic shift. High sensitivity of UV spectrometry allows detection of microgram level of gallic acid.

Table 1. Reported UV spectroscopy methods for analysis of gallic acid

S. No	Solvent Used	Detection Wavelength (nm)	Linearity/ Range (µg/ml)	LOD (µg/ml)	LOQ (µg/ml)	References
1.	Water	256.20	5-20	0.329	0.999	(Patel, 2019)
2.	Methanol	272	2-10	0.051	0.154	(Naaz et al., 2020)
3.	Water	220	4-20	0.147	0.447	(Roge and Shendarkar, 2018)
4.	Water	256	5-30	0.0405	0.119	(Sneha et al., 2022)
5.	Methanol	264	2-10	1.08	3.26	(Pancham and Patil, 2020)
6.	Methanol	273	5-30	0.014	0.045	(Pawar and Salunkhe, 2013)
7.	Methanol	227	10-50	-	-	(Kumar et al., 2010)
8.	Phosphate buffer saline	259	1 - 30	0.02	0.05	(Zanwar et al, 2022)

Infrared Spectroscopy (IR)

Infrared spectroscopy (IR) involves the absorption of different IR frequencies by a sample (Sumayya et al., 2016) to determine the chemical functional groups present in it (Hirun et al., 2012). Due to the unique fingerprinting capability of IR spectroscopy, when coupled with a chromatographic method like HPLC, it improves the analysis of complex mixtures. HPLC-FTIR is a rapid, accurate, sensitive and automated hyphenated technique having a wide range of application in combinatorial chemistry (Rakesh et al., 2014).

The infrared spectrum of gallic acid shows a stretching vibration of - OH between 3464.88 cm^{-1} (Sarria Villa et al., 2017) and a peak at 1705.92 cm^{-1} corresponding to a C=O group (Bishnoi et al., 2019). Two O-H bending bands can be observed at 1448.44 cm^{-1} and 868.87 cm^{-1}. C-H vibration is observed at 3167.12 cm^{-1} (Lubis et al., 2018). A C=C aromatic stretching vibration can be observed between 1636 cm^{-1}- 1526.55 cm^{-1} (Kamal et al., 2021). A band can be observed at 821.62 cm^{-1} corresponding to a vibration of the O-H deformation. Figure 3 shows major peaks of bonds/groups of gallic acid observed in the IR spectrum.

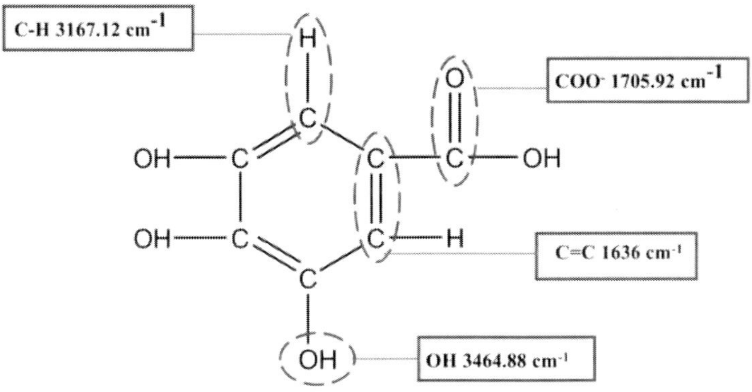

Figure 3. Major peaks of gallic acid obtained by IR spectroscopy.

Nuclear Magnetic Resonance (NMR)

NMR spectroscopy is one of the most powerful and useful analytical method which plays a major role in studying the radiofrequency radiated by nuclei in a magnetic field (Stark et al., 2016). The information obtained from NMR studies is vast, ranging from molecule structures to details of molecular interactions and the dynamics of molecular motions (Zhang et al., 2012). Being a non-destructive and non-invasive technique, proton NMR (^{1}H NMR) and carbon NMR (^{13}C NMR) are more valuable and useful tools for structural elucidation (Elyashberg, 2015) based on splitting patterns with the relative amounts of protons and carbons, respectively (Lavoie et al., 2016; Zhao et al., 2011). ^{1}H NMR of gallic acid exhibits peaks at δ 7.0313 ppm due to presence of aromatic protons at position 2 and 6 (Lubis et al., 2018). Phenolic protons at position 3 and 5 show peak at δ 9.15 ppm while the one at position 4 shows peak at δ 8.80 ppm. Due to anisotropy and electronegativity, the carboxylic acid proton at position 7 is deshielded to produce a peak at high δ value of 12.24 ppm. (Chen et al., 2018; López-Martínez et al., 2015).

In ^{13}C NMR studies of gallic acid, aromatic carbons at position 1 exhibits δ value at 119.87 ppm while both the carbons at position 2 and 6 exhibit δ value at 108.14 ppm (Ghareeb et al., 2019). Because of the presence of electronegative oxygen, the aromatic carbons attached to the -OH group are shifted downfield. Hence aromatic carbons at position 3, 4 and 5 show peaks at high δ values of 146.4652 ppm, 139.664 ppm and 146.4652 ppm, respectively. Aromatic carbon at position 7 is conjugated and attached to oxygen (C=O) as well as -OH resulting in highest δ value of 170.4923 ppm (Tukiran et al., 2016) (Figure 4).

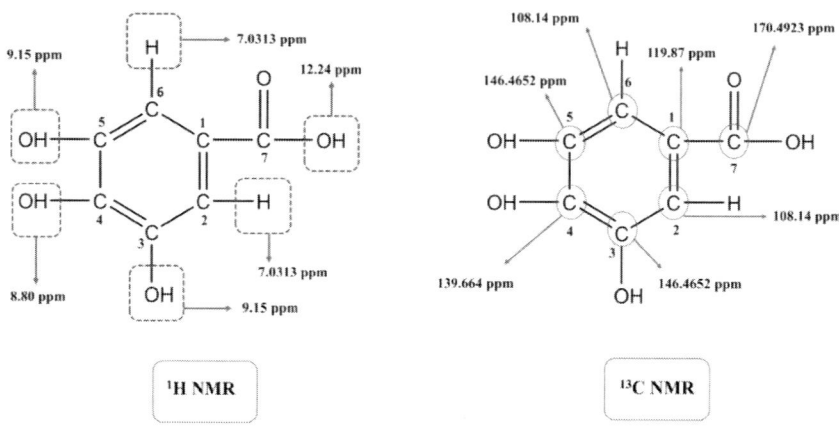

Figure 4. Major peaks of gallic acid obtained by ^{1}H NMR and ^{13}C NMR.

Mass Spectrometry (MS)

Since its discovery in 1912, mass spectrometry (MS) has been a commonly utilised instrumental method (van Bramer, 1998) for structural elucidation based on fragmentation pattern of analyte under study (Smith, 2013). The fundamental principle of MS is ionisation and fragmentation of sample molecules in the gas phase. When coupled with chromatographic techniques like gas chromatography (GC) (Al-Rubaye et al., 2017) or liquid chromatography (LC) (Desiderio & Fridland, 1984), mass spectroscopy allows simultaneous separation and quantification of compounds. Figure 5 exhibits a fragmentation pattern observed during mass analysis of gallic acid. The molecular ion for gallic acid is observed at m/z 170 and due to loss of -OH group, [(3,4,5-trihydroxyphenyl) methylidyne] oxidanium is formed with m/z value 153 which further undergoes fragmentation to lose C≡O group to form Benzene-1,2,3-triol with m/z value 125. A summary of reports that have been published on the analysis of gallic acid by Mass spectrometry is presented in Table 2.

Table 2. Reported mass spectrometry methods for analysis of gallic acid

S. No	Sample	Stationary Phase	Mobile Phase	Flow Rate (ml/min)	Retention Time (minutes)	MS Condition	m/z value	References
1	Dried Pomegranate flowers	ACE Excel 3 Super C18 column (100 mm × 2.1 mm, 3.0 μm)	Gradient using 0.1% Formic acid in water: 0.1% Formic acid in Acetonitrile	0.35	2.06	Capillary Voltage: Positive - 4000 V Negative - 3500 V Nozzle Voltage: 500 V Temperature and flow rate, 300°C and 7 L/min; sheath gas temperature and flow rate, 350°C and 12 L/min; Collision Energy (CE): 15 eV, 30 eV.	169.0140	(Yisimayili et al., 2019)
2	*Loropetalum chinense* (R. Brown) Oliv.	Welch C18 column (100 mm × 2.1 mm, 1.8 μm)	Gradient: Water with 0.1% Formic acid and Acetonitrile	0.25	1.75	Collision Energy: 35 eV Ion Spray Voltage: 4500V Turbo Spray Temperature:600°C Curtain gas: 25 psi	171.0284	(Chen et al., 2018)
3	*Phyllagathis rotundifolia*	Hypersil Gold RP C8 column (150 mm × 2.1 mm, 3 μm)	Gradient using 0.1% Formic acid in water: Acetonitrile	0.2	4.80	Ion spray voltage: -3.5 kV Capillary Temperature: 285°C	169.01428	(Tan et al., 2011)
4	*Terminalia arjuna*	Thermo Betasil C8 column (250 mm × 4.5 mm, 5μm)	Gradient using 0.1% Formic acid in water: Acetonitrile	0.5	9.9	Capillary Temperature: 350°C Nebulizer pressure: 35 psi. Flow rate: 10 L/min.	169.0141	(Singh et al., 2015)
5	Rat Plasma	Shimadzu Shim-pack VP- ODS C18 column (250 mm 9 × 2.0 mm, 5 μm)	0.1% Formic acid: Methanol (60:40 v/v)	0.2	-	Spray voltage: 3,800 V Capillary Temperature: 300°C, Sheath gas and Auxiliary gas (nitrogen) pressure at 27 and 6 arbitrary unites, respectively. Collision Energy: 20 V	169	(Song et al., 2010)

Table 2. (Continued)

S. No	Sample	Stationary Phase	Mobile Phase	Flow Rate (ml/min)	Retention Time (minutes)	MS Condition	m/z value	References
6	*Caesalpinia ferrea*	Phenomenex Synergi Hydro RP C18 column (250 mm × 10 mm, 10 μm)	Gradient: 0.1 Acetic acid in Water; 0.1 Acetic acid in Methanol	0.8	11.72	Capillary Voltage: −20 kV Spray Voltage: −5kV Capillary Temperature: 275°C	169	(Wyrepkowski et al., 2014)
7	*Myrcia*	-	-	-	-	Capillary Voltage: 4 kV Gas temperature: 290°C Flow rate: 11 L min^{-1} Nebulizer pressure: 45 psi. Collision Energy: 35 eV	169	(Santos et al., 2018)
8	Yucca species	XDB C18 column (50 mm × 2.1 mm, 1.8 μm)	Gradient: Water with 0.1% Formic acid and Acetonitrile	0.8	22.4	Capillary Voltage: 4000 V Endplate Voltage: 500 V Nitrogen (nebulizing gas):35.0 psi Fragmentation amplitude: 135 eV.	171	(el Sayed et al., 2020)
9	*Quercus aegilops* Root	Eclipse XDB HPLC C18 column (150 mm × 4.6 mm, 5 μm)	Water: 0.01% Trifluoroacetic acid in methanol (30: 70 v/v)	0.70	1.59	Ion Spray Voltage: −4500 V Source Temperature: 550°C	168.8	(Amr et al., 2021)
10	*Rhodiola kirilowii* (Regel.)	Acquity UPLC BEH C18 column (50 mm × 2.1mm, 5 μm)	Methanol: Water (95:5 v/v)	0.35	0.41	Ion Source Temperature: 100°C Desolvation Temperature: 300°C	169	(Gryszczyńska et al., 2013)

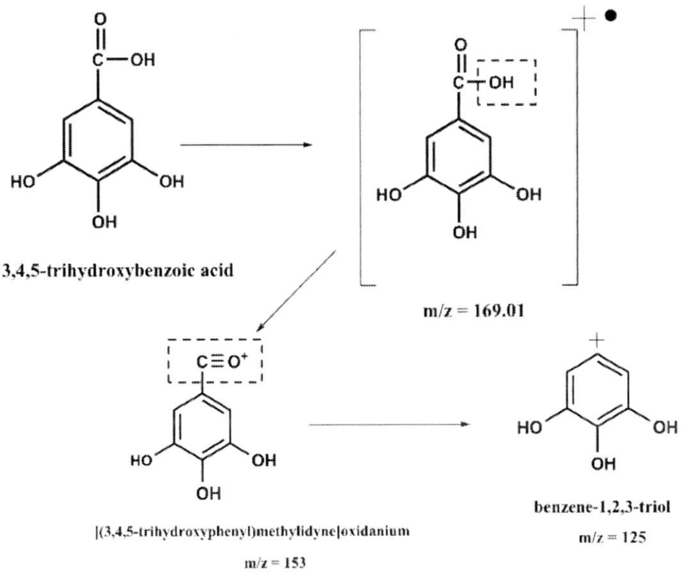

Figure 5. Schematic fragmentation pattern of gallic acid.

Thin Layer Chromatography (TLC)

TLC is the simplest chromatographic technique majorly used for qualitative analysis. It helps to identify the number of components present in a sample and can be evaluated in terms of R_f value (retardation factor) (Lade et al.).

In TLC, a thin coating of an adsorbent, such as silica gel, alumina, or cellulose, is used as the stationary phase on a flat support (Jumde and Gurnule, 2019) and sample is transported across the stationary phase by the mobile phase (Santiago and Strobel, 2013) through capillary action (Bele et al., 2011). TLC, like other chromatographic techniques, separates complicated mixtures on the basis of the varying affinities of the analyte for the mobile and stationary phases (Kowalska and Sajewicz, 2022). As silica gel is polar due to presence of Silanol group, TLC is majorly normal phase chromatography with non-polar solvents such as mobile phase which elutes non-polar compounds and retards polar compounds (Tiwari and Talreja, 2022). The most suitable mobile phase is the one that elutes the compounds with R_f values between 0.15 - 0.85. TLC is method of choice for preliminary investigation as it is a fast, easy and inexpensive analytical technique because it does not require sophisticated instrumentation and the amount of mobile phase required per analysis is very less (Sherma, 2006).

Table 3. Reported TLC methods for analysis of gallic acid

S. No	Stationary Phase	Mobile Phase (v/v)	Sample Volume applied	Distance travelled (cm)	Saturation time (min)	Wavelength (nm)	Rf	References
1	Merck Silica gel 60 GF 254	Hexane: Ethyl acetate: Formic acid (10: 5: 1)	5 μL	1.5	-	366	0.27	(Myadagbadam et al., 2022)
2	Precoated Silica gel 60 GF 254	Chloroform: Methanol: Formic acid (7: 2: 1)	5μL	8	10	254	-	(Sreedevi and Vijayalakshmi, 2018)
3	Silica gel GF 254	Toluene: Ethyl acetate: Formic acid (6: 6: 1)	20 mg	-	-	254	0.44	(Genwali et al., 2013)
4	Plastic backed neutral aluminum oxide GF 254	Chloroform: Ethyl acetate: Formic acid (5: 4: 1)	5 μl	-	-	522	0.4	(Hachula et al., 2004)
5	Precoated Silica gel GF 254	Toluene: Ethyl acetate: Methanol: Formic acid (6: 6: 0.4: 1.6)	-	-	10	275	0.82	(Koparde et al., 2020)
6	Precoated Silica gel GF 254	Chloroform: Methanol (9: 1)	20 μl	-	-	254	0.11 ± 0.009	(Saxena et al., 2017)
7	Precoated Silica gel GF 254	Toluene: Acetone: Formic acid (3: 6: 1)	-	-	-	254	0.88	(Sarria Villa et al., 2017b)

Many literatures are available for estimation of gallic acid done by TLC. Table 3 depicts a summary of published reports on quantification of gallic acid by TLC. As per reported methods for TLC studies of gallic acid, ethyl acetate in combination with hexane or chloroform are widely used as mobile phase for estimation of gallic acid. As gallic acid is acidic in nature, mobile phase modifiers such as formic acid are used in small volumes to prevent tailing.

High Performance Thin Layer Chromatography (HPTLC)

High Performance Thin Layer Chromatography (HPTLC) referred to as flat-bed chromatography or Planar Chromatography (Ramu and Chittela, 2018), is a highly advanced and automated type of thin layer chromatography (TLC) which offers sensitivity, accuracy and reproducibility equivalent to other analytical techniques (Srivastava MM). In contrast to TLC, this technique works well for both qualitative and quantitative analysis (Bhavya shree, 2017). A densitometer is used to scan separation tracks using visible or ultraviolet laser beams for quantification of separated bands at nanogram and picogram level (Pandhare et al., 2021). HPTLC offers a wide range of solvents as mobile phases because it is not limited by UV cut-off or purity of solvent used for sample preparation or used as mobile phase components. Additionally, a diverse range of stationary phases such as silica gel for normal phase and C8, C18 for reversed phase chromatography can be used (Rashmin et al., 2012). The possibility of interference or contamination from previous analysis is negligible as fresh stationary and mobile phases are used for each analysis (Attimarad et al., 2011). A summary of reported literature that has been published on the analysis of gallic acid by HPTLC is shown in Table 4.

The selection of the mobile phase is one of the most important parameters to achieve efficiency during separation in HPTLC. It is based on the compound's solubility with the solvent and the difference in the affinity for the mobile phase versus the stationary phase. The eluting power of solvents is directly proportional to its polarity. The less polar compounds can be readily eluted with low polarity solvents, while highly polar compounds require higher polarity solvents. Ethyl acetate (polarity index 4.4) is relatively more polar than other solvents such as toluene (polarity index 2.4) and chloroform (polarity index 4.1), its high volume resulting in high R_f values of gallic acid. Formic acid and acetic acid are used as mobile phase modifiers to prevent tailing.

Table 4. Reported HPTLC methods for analysis of gallic acid

S. No.	Stationary Phase	Mobile Phase (v/v)	Saturation Time (minutes)	Sample Volume applied	Distance travelled (mm)	Wavelength (nm)	Bandwidth (mm)	Rf	Range	LOD	LOQ	References
1	Pre coated silica gel 60 F254	Toluene: Ethyl acetate: Formic acid: Methanol (3: 3: 0.8: 0.2)	15	-	-	280	-	0.56	1.25 – 5.00	125 ng spot^{-1}	-	(Jeganathan and Kannan, 2008)
2	Aluminum plates Pre coated silica gel 60 F254	Toluene: Ethyl acetate: Formic acid (6: 6: 1)	10 - 30	-	-	297	6	0.41	100-500 ng/spot	0.67 ng per band	2.02 ng per band	(Veer et al., 2014)
3	Pre coated silica gel 60 F254	Toluene: Ethyl acetate: Methanol: Formic acid (6: 6: 0.4: 1.6)	10	-	-	275	8	0.8	0.2-2 ug/ml	-	-	(Koparde et al., 2020)
4	Pre coated silica gel 60 F254	Toluene: Acetone: Glacial acetic acid (3:1:2)	30	-	-	254	7	0.29	-	-	-	(Kumar et al., 2010)
5	Pre coated silica gel 60 F254	Toluene: Ethyl acetate: Methanol: Acetic acid: Formic acid (10.4: 4: 0.4: 0.3)	15	-	-	456	6	0.41	80-240 ng band^{-1}	6.05 ng band^{-1}	18.35 ng band^{-1}	(Potawale et al., 2022)
6	Pre coated silica gel 60 F254	Toluene: Ethyl acetate: Formic acid (5: 7: 1)	8	4μl	70	278	6	0.45	0.33 – 1.43μg	0.69ng	2.1ng	(Karthika et al., 2019)
7	Pre coated silica gel 60 F254	Toluene: Ethyl acetate: Formic acid: Methanol (3: 3: 0.8: 0.2)	-	2μl	90	254 -366	5	0.49	-	-	-	(Narayanan and Marimuthu alias Antonysamy, 2016)

S. No.	Stationary Phase	Mobile Phase (v/v)	Saturation Time (minutes)	Sample Volume applied	Distance travelled (mm)	Wavelength (nm)	Bandwidth (mm)	Rf	Range	LOD	LOQ	References
8	Pre coated silica gel 60 F254	Chloroform: Ethyl acetate: Formic acid: Methanol (7.5: 6: 0.5: 0.5)	-	1 μg per spot		322	8	0.25	3 – 10 μg per spot	0.5	16	(Chavan et al., 2015)
9	Pre coated silica gel 60 F254	Toluene: Ethyl acetate: Formic acid: Methanol (5.6: 2.2: 1.2: 1.0)	20	2-15 μl	90	327	8	0.29	250-550 ng/spot	14.770	44.759	(Vyas and Patel, 2016)
10	Pre coated silica gel 60 F254	Chloroform: Ethyl acetate: Formic acid (7.5:6: 0.5)	-	100 ng/spot	70	254	8	0.26	100–700	33.33 ng spot^{-1}	100 ng spot^{-1}	(Sonawane et al., 2011)
11	Pre coated silica gel 60 F254	Toluene: Ethyl acetate: Methanol: Glacial acetic acid (4: 4: 5:0.5)	30	2-10 μg/ spot	80	254	6	0.81	10-100μg/μl	0.114	0.345	(Kilaje and Shirsat, 2021)
12	Pre coated silica gel 60 F254	Toluene: Ethyl acetate: Methanol (5: 2: 3)	20	-	80	273	6	0.29	400– 1200 ng/band	100 ng/band	300 ng/band	(Rk et al.,2020)
13	Pre coated silica gel 60 F254	Toluene: Ethyl acetate: Methanol: Formic acid: (3: 3: 1: 0.4)	15	5.0 μl	80	280	7.2	0.21	100-3000 ng/spot	130	396.8	(Iman Tagelsir Abdalla Mohamed et al., 2020)
14	Pre coated silica gel 60 F254	Toluene: Ethyl acetate: Formic acid (4.5: 3: 0.2)	-	10 μl	80	366	-	0.40	600 to 1500 ng/spot	78.87	262.91	(Thakker et al., 2011)
15	Pre coated silica gel 60 F254	Chloroform: Methanol (7: 3)	-	10 μl		292	10	-	100 - 300	9	26	(Kumar et al., 2020)

Table 4. (Continued)

S. No.	Stationary Phase	Mobile Phase (v/v)	Saturation Time (minutes)	Sample Volume applied	Distance travelled (mm)	Wavelength (nm)	Bandwidth (mm)	Rf	Range	LOD	LOQ	References
16	Pre coated silica gel 60 F254	Ethyl acetate: Methanol: Formic acid (8: 2: 1).	20	1 μl	70	280	8	0.78	100 – 400 ng/spot	0.8018	2.4299	(Pradeep Patil et al.,2012)
17	Pre coated silica gel 60 F254	Ethyl acetate: Acetone: Water: Formic acid (10: 6: 2: 2)	15-20	-	-	254	6	0.83	100 – 700 ng/spot	7.43 ng/mL	18.20 ng/mL	(Alam et al., 2019)
18	Pre coated silica gel 60 F254	Toluene: Ethyl acetate: formic acid: methanol (3: 3: 0.6: 0.4)	15	-	80	280	6	0.51	0.4 - 2.0 mg/band	0.103 mg /spot	0.312 mg/spot	(Patel et al.,2010)
19	Pre coated silica gel 60 F254	Toluene: Ethyl acetate: Methanol: Formic acid (5: 4: 0.5: 0.5)	30	10 μl	80	254	8	0.33	200-1200 ng/ spot	75.24 ng/spot	227.99 ng/spot	(Alama and K. Mishra, 2020)
20	Pre coated silica gel 60 F254	Ethyl Formate: Toluene: Formic acid: Water (20: 1: 2.6: 0.5)	20	-	80	254 - 366	5	0.75	100 - 700	38.3	116.1	(Prashar and Patel, 2020)
21	Pre coated silica gel 60 F254	Toluene: Ethyl acetate: Formic acid (5: 5: 1)	60	-	-	254	-	0.91	200–1000μg/ml	4.71μg/ml	14.29μg/ml	(Gupta, 2016)
22	Pre coated silica gel 60 F254	Ethyl acetate: Toluene: Formic acid (8: 2: 0.3)	25	0.8 mg/ml	-	254	7	0.68	1.6 - 8 μg	0.12	0.365	(Chothani, 2014)
23	Pre coated silica gel 60 F254	Toluene: Ethyl Acetate: Formic Acid (7:5:1)	30	-	60	254	7	0.39	-	-	-	(Research Library and Manavalan, 2010)

S. No.	Stationary Phase	Mobile Phase (v/v)	Saturation Time (minutes)	Sample Volume applied	Distance travelled (mm)	Wavelength (nm)	Bandwidth (mm)	Rf	Range	LOD	LOQ	References
24	Pre coated silica gel 60 F254	Toluene: Ethyl acetate: Formic acid (5: 2.5: 0.5)	15	-	-	366	6	0.24	0.6-1.3	0.6	0.7	(Vasim et al., 2016)
25	Pre coated silica gel 60 F254	Toluene: Ethyl acetate: Methanol: Formic acid (4.9: 4.1: 2: 0.5)	10 - 20	-	80	300	6	0.22	200-1200 ng/band	-	-	(Thomas et al., 2020)
26	Pre coated silica gel 60 F254	Toluene: Ethyl acetate: Formic acid: Methanol (12: 9: 4: 0.5)	-	10 µl	80	273	6	0.4	-	-	-	(Siddiqui et al., 2014)
27	Pre coated silica gel 60 F254	Toluene: Ethyl acetate: Formic acid (4: 6: 1)	-	16–56 ng/ml	80	280	6	0.60	160 - 500	40	80	(Dhalwal et al., 2008)
28	Pre coated silica gel 60 F254	Toluene: Ethyl acetate: Formic acid (3: 2: 0.4)	-	10 µl	80	280	5	0.37	150–750	50 ng	150 ng	(Pathak et al., 2004)

High Performance Liquid Chromatography (HPLC)

High Performance Liquid Chromatography, also referred as High-Pressure Liquid Chromatography (Chawla and Kr. Chaudhary, 2019), is an analytical technique used to separate, recognise, and quantify each component of a mixture (Thammana, 2016). HPLC offers speed, sensitivity and reproducibility making it method of choice for both qualitative and quantitative analysis. (Baloch and Yang, 2021). When it is coupled with mass spectrometry (MS), it offers great capabilities in separation and mass analysis providing accurate data on sample composition (Attimarad and Alnajjar, 2013).

The primary components of HPLC are stationary phase (column holding compactly packed adsorbent), a mobile phase (solvent moving through the column), pump (to flow the mobile phase) and a detector that displays the retention time of the molecules (Malviya and Sharma, 2010). Separation occurs depending on the affinity of compounds with either stationary phase or with mobile phase (Sabir et al., 2016). Compounds with lesser affinity with stationary phase, tend to elute out more quickly along with the mobile phase and the compound having higher affinity with stationary phase will retain in the column for a longer period of time having higher retention time (Vidushi et al., 2017). Gallic acid being polar in nature; it gets easily soluble in polar solvent and can be analysed by reversed phase HPLC. It has an absorption maximum at 272.5 nm hence it is detected by a UV detector (Naaz et al., 2020). Many literatures are available for estimation of gallic acid by HPLC. Table 5 depicts a summary of published reports on quantification of gallic acid by HPLC.

Mobile phases commonly used in reported reversed-phase HPLC for gallic acid are hydro-organic mixtures including methanol and acetonitrile or combinations of these. Other mobile-phase modifiers such as Acetic acid (CH_3COOH), Orthophosphoric acid (H_3PO_4), Trifluoroacetic acid (TFA) ($C_2HF_3O_2$) have also been used for minor selectivity adjustments; however, they are not common or rarely preferred due to their high back pressure limitations and also high background UV absorbance. Acetonitrile is frequently used as a solvent for reversed-phase separations because of its extremely low UV cut-off 190 nm. In reversed phase chromatography, silica-based stationary phases are common, but alternative polymer-based adsorbents, such as styrene-divinylbenzene copolymer, are also used. Chemically bonded octadecyl silane (ODS) group, an alkane with 18 carbon atoms, is a well- known stationary phase for reverse phase HPLC separations.

Table 5. Reported HPLC methods for analysis of gallic acid

S. No	Stationary Phase	Mobile Phase (v/v)	Detection Wavelength (nm)	Retention time (min)	Linearity (µg/mL)	LOD (µg/mL)	LOQ (µg/mL)	References
1.	Phenomenex Luna C18 (250 x 4.6mm, 5 µm)	Water: Acetonitrile (80: 20)	272	3.6	0.5-50	0.0178	0.05399	(Kardani et al., 2013)
2.	Phenomenex Luna C18 (250 x 4.6mm, 5 µm)	Water pH 3 maintained by O-phosphoric acid: Acetonitrile (80: 20)	272	5.343	10-100	0.65	2.5	(Singh et al., 2019)
3.	LiChrospher 100 RP-18 column. (125 × 4 mm, 5 µm)	0.01 M Potassium dihydrogen phosphate: Acetonitrile (85:15), pH 3.2	280	1.98	6.25-50	0.75	2.50	(Zakaria et al., 2014)
4.	Phenomenex Gemini NX C18 column (250 × 4.6mm, 5µm)	Gradient using 0.05% Phosphoric acid in Water and Methanol	271	8.5	2.5 -15	1.22	1.82	(Fernandes et al., 2015)
5.	Hypersil Gold C18 column (250 x 4.6mm, 5 µm)	Gradient using 1% aqueous Acetic acid and Methanol	278	5.42	5 - 25	0.179	0.599	(Nour et al., 2013)
6.	Diamonsil C18 (250 x 4.6mm, 5 µm)	Methanol and 0.1% Phosphoric acid	260	-	7.375-236	-	-	(Sun et al., 2021)
7.	Waters' symmetry C-18 column (250 x 4.6mm, 5 µm)	Methanol: Ethyl acetate: Water (25:5:70)	270	1.94	0.01-0.06	-	-	(Sawant and Chavan, 2013)
8.	Enable C18G (250 x 4.6mm, 5 µm)	0.1% Acetic acid in Water: Methanol (10:90)	272	2.632	4 -24	0.75	1.65	(Shalavadi et al., 2019)
9.	Octadecyl silane (250 x 4.6 mm, 5 µm)	Acetonitrile: (0.01% O-phosphoric acid in water) (20: 80)	270	2.81	20 - 120	6.13	18.57	(Perumal, 2009)
10	Phenomax Luna C18 (250 × 4.6 mm, 5 µm)	Acetonitrile: Water (70:30)	255	2.77	40-100	10	20	(Krupa and Vishnu, 2021)
11	Grace C18 column (250 × 4.6 mm, 5 µm)	Acetonitrile: Water (60:40) (0.05% OPA)	210	5.90	0.5–2.5	0.0128	0.0389	(Bhishnurkar et al., 2020)

Table 5. (Continued)

S. No	Stationary Phase	Mobile Phase (v/v)	Detection Wavelength (nm)	Retention time (min)	Linearity (μg/mL)	LOD (μg/mL)	LOQ (μg/mL)	References
12	Phenomax Luna C18 (250 × 4.6 mm, 5 μm)	-	264	3.410	2 - 20	0.478	1.456	(Prakash Gupta and Garg, 2014)
13	Agilent Zorbax Eclipse XDB-C18 column (150× 4.6 mm, 5 μm)	Gradient using 10-mM Phosphoric acid solution and Methanol	214	5.452	5-100	0.99	3.32	(Türköz Acar et al., 2018)
14	C_{18} column (250 × 4.6 mm, 5 μm)	Methanol: 0.2% OPA in Water (20:80)	220	4.864	0.25-20	0.062	0.188	(Jain and Prabhu, 2022)
15	Platinum EPS C18 (53 × 7mm, 3 μm)	Water: 0.05% Trifluoroacetic acid in Acetonitrile (87:13)	210	7	1-100	0.2	1.0	(Theppakorn and Wongsakul, 2012)
16	VertiSepTM pHendure C18 (250 × 4.6 mm, 5 μm)	Acetonitrile: Formic acid (0.01%), (15:85), pH 2.5	275	3.86	0.25 - 20	0.01	0.04	(Pramote et al., 2018)
17	Phenomenex Hypersil C18 (DDS) column (250 × 4.6 mm, 5 μm)	Acetonitrile: Aqueous Acetic acid (10:1)	280	5.45	5-70	0.13	0.43	(Mohamed et al., 2021)
18	Nucleodur C18 (250 × 4.6 mm, 5 μm)	Buffer: Methanol: Acetonitrile (50: 30: 20)	263	2.15	1-10	0.130961	0.39685	(Gupta et al., 2019)
19	C18 reverse phase column (250 x 4.6mm, 5 μm)	0.1% Glacial acetic acid: Acetonitrile (70:30)	254	3.269	5-500	1.53	5.12	(Amir et al., 2013)
20	Kingsorb C18 (150 × 4.6 mm, 5 μm)	0.1% (v/v) Orthophosphoric acid in Water (0.1% (v/v): Orthophosphoric acid in Methanol	210	-	0.3 - 60.7	0.43	1.52	(Wang et al., 2003)
21	Shim-pack GIST C18 column (250 x 4.6mm, 5 μm)	Gradient using 1% Acetic acid and Acetonitrile	272	6.7	5 - 50	0.45	1.36	(Rajendra Kshirsagar et al., 2020)

These characteristics are appropriate for estimation for polyphenols such as gallic acid with improved selectivity and better resolution. The choice of selection of mobile phase composition directly or indirectly influences sample matrix retention in the column as well as peak resolution.

Conclusion

The selection of analytical techniques for estimation of analytes is very crucial as it affects specificity and sensitivity of the quantification. Structure of the compound and its chemistry plays an important role for the analyte to be analysed by a given analytical method. Qualitative and quantitative determination of gallic acid can be carried out by a number of analytical procedures. UV spectrometry is a simple method for analysis of pure gallic acid but it is not a specific method. Many IR and NMR spectroscopic methods are reported for identification of functional groups and position of protons/carbons of gallic acid, respectively. The relative position of peaks is affected by the type of sampling techniques used for these spectroscopies. Both HPLC and HPTLC offer specificity, sensitivity, accuracy and reproducibility for quantification. When these techniques are coupled with mass spectroscopy, the sensitivity is increased by many folds.

References

Abbasi, S., Daneshfar, A., Hamdghadareh, S. and Farmany, A. (2011). Quantification of Sub-Nanomolar Levels of Gallic Acid by Adsorptive Stripping Voltammetry. *International Journal of Electrochemical Science*, 4843-4852.

Al- Rubaye, A.F., Hameed, I.H and Kadhim, M.J. (2017). A Review : Uses of Gas Chromatography-Mass Spectrometry (GC-MS) Technique for Analysis of Bioactive Natural Compounds of somee plants. *International Journal of Toxicological and Pharmacological Research,* 9(01).

Alam, P., Kamal, Y. T., Alqasoumi, S. I., Foudah, A. I., Alqarni, M. H. and Yusufoglu, H. S. (2019). HPTLC method for simultaneous determination of ascorbic acid and gallic acid biomarker from freeze dry pomegranate juice and herbal formulation. *Saudi Pharmaceutical Journal*, 975-980.

Alama, G. and K. Mishra, A. (2020). Simultaneous Estimation of Gallic Acid and Embelin by Validated HPTLC Method in Three Marketed Formulations and in-House Formulated *Manibhadra Yoga* : A Polyherbal Ayurvedic Formulation. *Oriental Journal of Chemistry*, 120-126.

Aleixandre, J.L., Aleixandre-Tudo, J.L., Bolanos-Pizzaro, M. and Aleixandre-Benavent, R (2013). Mapping the scientific research on wine and health (2001-2011). *Journal of Agricultural and Food Chemistry.*

Amir, M., Mujeeb, M., Ahmad, S., Akhtar, M., Yt, K. and Ashraf, K. (2013) Simultaneous Quantitative HPLC Analysis of Ascorbic Acid, Gallic Acid, And Catechin In *Punica Granatum, Tamarindus Indica* And *Prunus Domestica* Using Box-Behnken Statistical Design. *World Journal of Pharmaceutical Research*, 1403-1416.

Amr, A. S., Ahmad, M. N., Zahra, J. A. and Abdullah, M. A. (2021). HPLC/MS-MS. Identification of Oak Quercus aegilops Root Tannins. *Journal of Chemistry*, 1-10.

Aruoma, O., Murcia, A., Butler, J. and Halliwellt, B. (1993). Evaluation of the Antioxidant and Prooxidant Actions of Gallic Acid and Its Derivatives. *Journal of Agricultural and Food Chemistry*, 1880-1885.

Attimarad, M. and Alnajjar, A. (2013). A conventional HPLC-MS method for the simultaneous determination of ofloxacin and cefixime in plasma : Development and validation. *Journal of Basic* and *Clinical Pharmacy*, 36-41.

Attimarad, M., Mueen, Ahmed., K. K., Aldhubaib, B. E. and Harsha, S. (2011). High-performance thin layer chromatography : A powerful analytical technique in pharmaceutical drug discovery. *Pharmaceutical Methods*, 71-75.

Baloch, S. and Yang, Y. (2021). Review on Methods and Applications of High-Performance Liquid Chromatography. *Journal of Food Processing and Technology*, 858.

Bele, A. A. and Khale, A. (2011). An Overview on Thin Layer Chromatography. *International Journal of Pharmaceutical Sciences and Research*, 256-267.

Bhavya shree, B. and Sonia, K. (2017). HPTLC Method Development and Validation : An Overview. *Journal of Pharmaceutical Sciences and Research,* 652-657.

Bhishnurkar, P., Deo, S. S., Inam, F. S., Mahmood, S. H., Taher, D. and Lambat, T. L. (2020). Simultaneous determination of β-sitosterol and gallic acid in *Nigella Sativa* seeds using reverse phase high performance liquid chromatography. *SN Applied Sciences*, 1873.

Bishnoi, R. S., Kumar, M., Shukla, A. K. and Jain, C. P. (2019). Comparative fingerprint and extraction yield of *Prosopis cineraria (Lin.)* Druce. Leaves with phenolic compounds (Gallic acid) & flavonoids (Rutin). *Journal of Drug Delivery and Therapeutics,* 560-568.

Boulet, J. C., Ducasse, M. A. and Cheynier, V. (2017). Ultraviolet spectroscopy study of phenolic substances and other major compounds in red wines : relationship between astringency and the concentration of phenolic substances. *Australian Journal* of *Grape* and *Wine* Research, 193-199.

Chavan, A. K., Nirmal, S. A. and Pattan, S. R. (2015). Development and Validation of HPTLC Method to Detect Curcumin and Gallic Acid in Polyherbal Microencapsulated Formulation. *Journal* of *Liquid Chromatography* & *Related Technologies*, 1213–1217.

Chawla, G. and Chaudhary, K. K. (2019). A review of HPLC technique covering its pharmaceutical, environmental, forensic, clinical and other applications. *International Journal of Pharmaceutical Chemistry and Analysis*, 27-39.

Chen, H., Li, M., Zhang, C., Du, W., Shao, H., Feng, Y., Zhang, W. and Yang, S. (2018). Isolation and identification of the anti-oxidant constituents from *Loropetalum chinense (R. Brown)* oliv. Based on UHPLC–Q-TOF-MS/MS. *Molecules*, 23.

Chothani, D. L. and Patel, N. M. (2014). Quantification of gallic acid in fruit and leaves of *Careya arborea* by High performance thin layer chromatography (HPTLC). *Journal of Investigational Biochemistry*, 138-142.

Choubey, S., Goyal, S., Varughese, L. R., Kumar, V., Sharma, A. K. and Beniwal, V. (2018). Probing Gallic Acid for Its Broad-Spectrum Applications. *Mini-Reviews in Medicinal Chemistry*, 1283-1293.

Daglia, M., Lorenzo, A., Nabavi, S., Talas, Z. and Nabavi, S. (2014). Polyphenols : Well Beyond the Antioxidant Capacity : Gallic Acid and Related Compounds as Neuroprotective Agents : You are What You Eat! *Current Pharmaceutical Biotechnology*, 362-372.

Daneshfar, A., Ghaziaskar, H. S. and Homayoun, N. (2008). Solubility of Gallic Acid in Methanol, Ethanol, Water, and Ethyl Acetate. *Journal of Chemical & Engineering Data*, 776-778.

Desiderio, D.M., and Fridland, G.H. (1984). A review of combined liquid chromatography and mass spectrometry. *Journal of Liquid Chromatography*, 7, 317-351.

Dhalwal, K., Shinde, V. M., Biradar, Y. S. and Mahadik, K. R. (2008). Simultaneous quantification of bergenin, catechin, and gallic acid from *Bergenia ciliata* and *Bergenia ligulata* by using thin-layer chromatography. *Journal of Food Composition and Analysis*, 496-500.

El Sayed, A. M., Basam, S. M., El-Naggar, E. M. bellah A., Marzouk, H. S. and El-Hawary, S. (2020). LC–MS/MS and GC–MS profiling as well as the antimicrobial effect of leaves of selected Yucca species introduced to Egypt. *Scientific Reports*, 10.

Elyashberg, M. (2015). Identification and structure elucidation by NMR spectroscopy. *Trends in Analytical Chemistry*, 1-18.

Eslami, A. C., Pasanphan, W., Wagner, B. A. and Buettner, G. R. (2010). Free radicals produced by the oxidation of gallic acid : An electron paramagnetic resonance study. *Chemistry Central Journal*, 1-4.

Fernandes, F. H. A., de Batista, R. S. A., de Medeiros, F. D., Santos, F. S. and Medeiros, A. C. D. (2015). Development of a rapid and simple HPLC-UV method for determination of gallic acid in Schinopsis brasiliensis. *Revista Brasileira de Farmacognosia*, 208-211.

Fernandes, F. H. A. and Salgado, H. R. N. (2016). Gallic Acid : Review of the Methods of Determination and Quantification. *Critical Reviews in Analytical Chemistry*, 257-265.

Fischer, E. (1914). Synthesis of Depsides, Lichen-substances and Tannins. *Journal of the American Chemical Society*, 1170-1201.

Fitzpatrick, L. R. and Woldemariam, T. (2017). Small-Molecule Drugs for the Treatment of Inflammatory Bowel Disease. *Comprehensive Medicinal Chemistry III*. Elsevier Inc., 495-510.

Fiuza, S. M., Gomes, C., Teixeira, L. J., Girão Da Cruz, M. T., Cordeiro, M. N. D. S., Milhazes, N., Borges, F. and Marques, M. P. M. (2004). Phenolic acid derivatives with potential anticancer properties - A structure-activity relationship study. Part 1: Methyl, propyl and octyl esters of caffeic and gallic acids. *Bioorganic and Medicinal Chemistry*, 3581-3589.

Gandhimathi, R., Vijayaraj, S. and Jyothirmaie, M. P. (2012). Analytical Process of Drugs by Ultraviolet (UV) Spectroscopy-A Review. *International Journal of Pharmaceutical Research and Analysis*, 72-78.

Genwali, G. R., Acharya, P. P. and Rajbhandari, M. (2013). Isolation of Gallic Acid and Estimation of Total Phenolic Content in Some Medicinal Plants and Their Antioxidant Activity. *Nepal Journal of Science and Technology*, 95-102.

Ghareeb, M. A., Sobeh, M., El-Maadawy, W. H., Mohammed, H. S., Khalil, H., Botros, S. and Wink, M. (2019). Supplementary Material Chemical Profiling of Polyphenolics in Eucalyptus globulus and Evaluation of Its Hepato-Renal Protective Potential against Cyclophosphamide Induced Toxicity in Mice, *Antioxidants*, 1-4.

Goldberg, I. and Rokem, J. S. (2009). *Organic and Fatty Acid Production, Microbial*. Elsevier Inc., 421-442.

Gryszczynska, A., Opala, B., Lowicki, Z., Krajewska-Patan, A., Buchwald, W., Czerny, B., Mielcarek, S., Boroń, D., Bogacz, A. and Mrozikiewicz, P. M. (2013). Determination of chlorogenic and gallic acids by UPLC-MS/MS. *Herba Polonica*, 7–16.

Gupta, M. and Kaur, A. (2019). Method Development and Validation for Simultaneous Estimation of Resveratrol and Gallic Acid by RP-HPLC. *International Journal of Pharmaceutical, Chemical and Biological Sciences*, 19-26.

Gupta, R. (2016). Determination of Gallic Acid and β - Sitosterol in Poly-Herbal Formulation by HPTLC. *Pharmacy and Pharmacology International Journal*, 373-379.

Gupta, S. P. and Garg, G. (2014). Quantitative estimation of gallic acid and tannic acid in Bhuvnesvara Vati by RP-HPLC. *Der Pharmacia Lettre,* 31-36.

Hachuła, U., Anikiel, S. and Polowniak, M. (2004). Determination of gallic acid after thin-layer chromatographic and paper chromatographic separations. *Journal of Planar Chromatography*, 51–53.

Herrmann, K. M. and Weaver, L. M. (1999). The Shikimate Pathway, *Annual Review of Plant Physiology and Plant Molecular Biology*, 473-503.

Hirun, N., Dokmaisrijan, S. and Tantishaiyakul, V. (2012). Experimental FTIR and theoretical studies of gallic acid-acetonitrile clusters. *Spectrochimica Acta Part A: Molecular and Biomolecular Spectroscopy*, 93-100.

Jain, V. and Prabhu, K. P. (2022). Development of Novel Reverse Phase High Performance Liquid Chromatography Method for Simultaneous Estimation of Gallic Acid, Protocatechuic Acid, Vanillic Acid and Syringic Acid in Sugarcane Roots. *Indian Journal of Pharmaceutical Sciences*, 66-71.

Jeganathan, N. S. and Kannan, K. (2008). HPTLC Method for Estimation of Ellagic Acid and Gallic Acid in Triphala churanam Formulations. *Research Journal of Phytochemistry*, 1-9.

Jumde, M. H., Gurnule, W. B. (2019). An Overview of Thin Layer Chromatography. *International Journal of Management, Technology and Engineering*, 2846-2854.

Kamal, A. A. A., Mohamad, S., Mohamad, M., Wannahari, R., Sulaiman, A. Z., Ajit, A., Lim, J. W. and Zaudin, N. A. C. (2021). Extraction of gallic acid from *Chromolaena sp.* Using ultrasound-assisted extraction. *Engineering Journal*, 269-276.

Kardani, K., Gurav, N., Solanki, B., Patel, P. and Patel, B. (2013). RP-HPLC method development and validation of gallic acid in Polyherbal tablet formulation. *Journal of Applied Pharmaceutical Sciences*, 37-42.

Karthika, R. S., Hameed, A. S. and Meenu, M. T. (2019). International Journal of Ayurveda and Pharma Research HPTLC Estimation of Gallic Acid and Ellagic Acid in *Amrtottara kvatha* prepared in two ratios. *International Journal of Ayurveda and Pharma Research*, 1-9.

Kilaje, S. V. and Shirsat, M. K. (2021). Simultaneous Estimation of Ascorbic Acid and Gallic Acid in Triphala Ghrita Formulation by Hptlc, Volatiles & Essent. Oils. *Natural Volatiles and Essential Oils*, 3240-3249.

Koparde, A. A., Doijad, R. C., Patil, A. and Magdum, C. S. (2020). Isolation and Quantification of Gallic Acid from *Eulophia Ochreata* Lindl. By HPTLC. *International Journal of Pharmaceutical Sciences and Research*, 3859-3866.

Kowalska, T. and Sajewicz, M. (2022). Thin-Layer Chromatography (TLC) in Screening of Botanicals-its Versatile Potential and Selected Applications. *Molecules,* 27, 6607.

Krupa, G. and Vishnu, P. (2021). HPLC Method Development for Estimation Gallic acid and Ellagic acid in an Ayurvedic Anti Arthritic formulation Simhnad Guggul. *Tropical journal of Pharmaceutical and Life Sciences*, 1-12.

Kshrisagar, R. R., Vaidya, S. A. and Jain, V. (2020). Development and validation of a novel RP-HPLC method for the simultaneous quantification of ascorbic acid, gallic acid, ferulic acid, piperine, and thymol in a polyherbal formulation. *Indian Journal of Natural Products and Resources,* 307-311.

Kumar, A., Lakshman, K., Jayaveera, K. N., Tripathi, S. N. M. and Satish, K. (2010). Estimation of Gallic Acid, Rutin and Quercetin in *Terminalia chebula*. *Jordan Journal of Pharmaceutical Sciences*, 63-68.

Kumar, D., Madaan, R. and Kumar, S. (2020). Development of hptlc method for estimation of gallic acid and bergenin in *Actaea acuminata* roots. *Indian Journal of Pharmaceutical Education and Research*, 1169-1173.

Kumar, R. S., Kishan, R. J., Roa, V. K. and Kumanam, R. (2010). Simultaneous Spectrophotometric Estimation of Curcuminoids and Gallic Acid in Bulk Drug and Ayurvedic Polyherbal Tablet Dosage Form. *International Journal of Pharmaceutical Quality Assurance*, 56-59.

Lade, B. D., Patil, A. S., Paikrao, H. M. and Kale, A. S. (2014). A Comprehensive Working, Principles and Applications of Thin Layer Chromatography. *Research Journal of Pharmaceutical, Biological and Chemical Sciences*, 486-503.

Lavoie, S., Ouellet, M., Fleury, P. Y., Gauthier, C., Legault, J. and Pichette, A. (2015). Complete 1H and 13C NMR assignments of a series of pergalloylated tannins. *Magnetic Resonance in Chemistry*, 168–174.

Lide, D. R., Baysinger, G., Berger, L. I., Goldberg, R. N., Kehiaian, H. V., Kuchitsu, K., Roth, D. L., Zwillinger, D., *CRC Handbook of Chemistry and Physics*.

López-Martínez, L. M., Santacruz-Ortega, H., Navarro, R. E., Sotelo-Mundo, R. R. and González-Aguilar, G. A. (2015). A 1H NMR investigation of the interaction between phenolic acids found in mango (*Manguifera indica cv Ataulfo*) and papaya (*Carica papaya cv Maradol*) and 1,1-diphenyl-2-picrylhydrazyl (DPPH) free radicals. *PLoS One,* 10.

Lubis, M. Y., Marpaung, L., Nasution, M. P. and Simanjuntak, P. (2018). Gallic acid from pods of Jiringa (*Archidendron Jiringa* [Jack] I. C. Nielsen) and its antioxidant. *Asian Journal of Pharmaceutical and Clinical Research,* 114-117.

Luis, A. T., Buica, J., Nieuwoudt, A., H. H., Luis A, J. and Johannes Du Toit, W. (2017). Spectrophotometric Analysis of Phenolic Compounds in Grapes and Wines : A Review, *Journal of Agricultural and Food Chemistry,* 1-47.

Malviya, R. and Sharma, P. (2010). Polymer Modification and Applications View project extraction of okra mucilage. *Journal of Global Pharma Technology,* 22-26.

Manish Pal, S., Avneet, G. and Siddhraj, S. S. (2018). Gallic Acid: Pharmacogical Promising Lead Molecule: A Review. *International Journal of Pharmacognosy and Phytochemical Research,* 132–138.

Mohamed, I. T. A., Khan, W., Chester, K., Mohamed, A. H., Ahmad, S. and Ayoub, S. M. H. (2020). Simultaneous quantitative estimation of ellagic acid and gallic acid in Sudanese *Solanum dubium* seed by high performance thin layer chromatography (HPTLC). *GSC Biological and Pharmaceutical Sciences,* 54-61.

Mohamed, S. H., Youssef, A. F. A., Issa, M., Abdel Salam, H. S. and EL-Ansary, A. L. (2021). Validated HPLC method for quantitative analysis of gallic acid and Rutin in leaves of *Moringa oleifera* grown in Egypt. *Egyptian Journal of Chemistry,* 1583–1591.

Myadagbadam, U., Purevsuren, S., Chimedragchaa, C., Tserenkhand, G. and Norovnyam, R. and (2022). Standardization Study of Khurtsiin deed-6 Traditional Medicine. *Pharmacognosy Journal,* 610–621.

Naaz, H., Srikanth, P., RudrapaL, M. and Sarwa, K. K. (2020). Development and validation of UV spectrophotometric and RP-HPLC methods for the estimation of gallic acid in herbal formulation of amalaki. *Asian Journal of Chemistry,* 2469–2474.

Narayanan, J. and Marimuthu, J. (2016). HPTLC Fingerprint Profile (Phenolics) of Selected *Cyathea Species* from Western Ghats, South India. *Chinese Journal of Biology,* 1–7.

Nour, V., Trandafir, I. and Cosmulescu, S. (2013). HPLC determination of phenolic acids, flavonoids and juglone in walnut leaves. *Journal of Chromatographic Science,* 883–890.

Pancham, Y. P. and Patil, N. A. (2020). Validated UV-Spectrophotometric Method for Simultaneous Estimation of Curcumin and Gallic acid in bulk powder. *World Journal of Pharmacy and Pharmaceutical Sciences,* 1255-1266.

Pandhare, R. B., Kulkarni, R. N., Deshmukh, V. K., Mohite, P. B. and Pawar, A. R. (2021). High-performance thin layer chromatography : A powerful analytical technique in pharmaceutical drug discovery. *Journal of Pharmaceutical and Biological Sciences,* 7–14.

Patel, K. G., Patel, V. G., Patel, K. V. and Gandhi, T. R. Validated HPTLC Method for Quantitative Determination of Gallic Acid in Stem Bark of *Myrica esculenta* Buch. - Ham. Ex D. Don, Myricaceae. *Journal of AOAC International,* 1422-1427.

Patel, T. and Hinge, M. (2019). Development and Validation of UV- Spectrophotometric Method for Estimation of Gallic acid in herbal formulation. *International Journal of Recent Scientific Research,* 35391-35396.

Pathak, S. B., Niranjan, K., Padh, H. and Rajani, M. (2004). TLC densitometric method for the quantification of eugenol and gallic acid in clove. *Chromatographia,* 241–244.

Patil, V. P., Kurhade S. D., Devdhe S. J., Kale, S. H., Wakte P. S. (2012). Determination of Gallic Acid by HPTLC as Quality Control Parameter in Herbal Formulation : Triphala churna. *International Journal of Chemical and Analytical Science,* 1546–1549.

Pawar, N. P., Salunkhe, V. R. (2013). Development and Validation of UV Spectrophotometric Method For Simultaneous Estimation Of Rutin And Gallic Acid In Hydroalcoholic Extract Of Triphala churna, *International Journal of PharmTech Research,* 724-729.

Potawale, R., Parker, H. and Karwande, A. (2022). Validated HPTLC Method for Simultaneous Quantification of Gymnemagenin and Gallic acid in Herbal Dosage Form. *Journal of Advanced Scientific Research,* 136–141.

Pramote, B., Waranuch, N. and Kritsunankul, O. (2018). Simultaneous Determination of Gallic Acid and Catechins in Banana Peel Extract by Reversed-Phase High Performance Liquid Chromatography, *Naresuan University Journal : Science and Technology,* 189-200.

Prashar, Y. and Patel, N. (2020). High-performance thin-layer chromatography analysis of gallic acid and other phytoconstituents of methanolic extracts of *Myrica nagi* fruit. *Pharmacognosy Research,* 95-101.

Ramu, B. and Chittela, K. B. (2018). High Performance Thin Layer Chromatography and Its Role Pharmaceutical Industry : Review. *Open Science Journal of Bioscience and Bioengineering,* 29–34.

Rashmin, P., Mrunali, P., Nitin, D., Nidhi, D. and Bharat, P. (2012). HPTLC method development and validation : Stratrgy to minimize methodological failures. *Journal of Food and Drug Analysis.*

Roge, A. B. and Shendarkar, G. R. (2018). Development of Validated UV Spectrophotometric Stability indicating Method for Estimation of Gallic acid in bulk form. *International Journal of Pharmacy and Biological Sciences,* 169–176.

Rosas, E. C., Correa, L. B. and Henriques, M. D. G. (2019). Antiinflammatory Properties of Schinus terebinthifolius and Its Use in Arthritic Conditions. In : *Bioactive Food as Dietary Interventions for Arthritis and Related Inflammatory Diseases.* Elsevier, 489–505.

S, S.S., Lawarence, B., Babu, D.K., (2016). FTIR Analyis and reverse phase high performance chromatographic determination of Phenolic acids of *Hypnea musciformis* (wulfen) J.V. Lamouroux. *International Journal of Pharmaceutics.* 6, 97-104.

Sabir, A. M., Moloy, M., Bhasin, P. S. (2016). HPLC Method Development and Validation: A Review. *International Research Journal of Pharmacy,* 39–46.

Sameermahmood, Z., Raji, L., Saravanan, T., Vaidya, A., Mohan, V. and Balasubramanyam, M. (2010). Gallic acid protects RINm5F β-cells from glucolipotoxicity by its antiapoptotic and Insulin-secretagogue actions. *Phytotherapy Research,* 83-94.

Santiago, M. and Strobel, S. (2013). Thin layer chromatography. *Methods in Enzymology*. 303–324.
Santos, C. dos, Galaverna, R. S., Angolini, C. F. F., Nunes, V. V. A., de Almeida, L. F. R., Ruiz, A. L. T. G., de Carvalho, J. E., Duarte, R. M. T., Duarte, M. C. T. and Eberlin, M. N. (2018). Antioxidative, antiproliferative and antimicrobial activities of phenolic compounds from three *Myrcia* species. *Molecules*, 23.
Sarria Villa, R. A., Corredor, J. A. G. and Isabel P, M. (2017). Isolation of Catechin and Gallic Acid from Colombian Bark of *Pinus patula*. *Chemical Sciences Journal*, 1-11.
Sawant, N. R. and Chavan, A. R. (2013). Determination of Gallic acid from their Methanolic Extract of Punica granatum By HPLC Method. *International Journal of ChemTech Research*, 2598-2602.
Saxena, R., Sharma, R. and Nandy, B. C. (2017). Chromatographic Determination of Phenolic profile from *Punica Granatum* fruit peels. *International Research Journal of Pharmacy*, 61–65.
Shalavadi, M. H., Chandrashekhar, V. M. and Muchchandi, I. S. (2019). High-performance liquid chromatography analysis of gallic acid and kaempferol in chloroform and ethanol extract of *Cassia hirsuta* seeds. *International Journal of Green Pharmacy*, 236-241.
Sherma, J. (2006). Thin-Layer Chromatography. *Encyclopedia of Analytical Chemistry*, 1-14.
Siddiqui, S., Usmanghani, K., Zahoor, A., Sheikh, Z. A. and Khan, S. S. (2014). Quantitative Estimation of Gallic Acid as Biomarker in Lipitame Tablets by HPTLC Densitometry for Diabetic Dyslipidemia. *Chinese Medicine*, 170–178.
Singh, M. P., Gupta, A., Sisodia S. S. (2019). Qualitative Analysis of Gallic Acid by HPLC Method in Different Extracts of Terminalia *Bellerica Roxb.* Fruit. *Journal of Pharmaceutical Sciences*, 101-106.
Smith, R.W. (2013). Mass Spectrometry. In *Encyclopedia of Forensic Sciences: Second Edition*. 603-608.
Sonawane, S. D., Nirmal, S. A., Patil, A. N. and Pattan, S. R. (2011). Development and validation of HPTLC method to detect curcumin and Gallic acid in polyherbal formulation. *Journal of Liquid Chromatography and Related Technologies*, 2664–2673.
Song, R., Xu, L., Zhang, Z., Tian, Y., Xu, F. and Dong, H. (2010). Determination of gallic acid in Rat plasma by LC-MS-MS. *Chromatographia*, 1107–1111.
Sreedevi, P. and Vijayalakshmi, K. (2018). Determination of antioxidant capacity and gallic acid content in ethanolic extract of *Punica granatum L.* Leaf. *Asian Journal of Pharmaceutical and Clinical Research*, 319–323.
Stark, N. M., Yelle, D. J. and Agarwal, U. P. (2016). *Techniques for Characterizing Lignin*, Elsevier. Inc, 49-65.
Sun, Y., Liu, J., Bayertai, Tang, S. and Zhou, X. (2021). Analysis of gallic acid and ellagic acid in leaves of *Elaeagnus angustifolia L.* From different habitats and times in Xinjiang by HPLC with cluster analysis. *Acta Chromatographica*, 195–201.
Tan, H. P., Ling, S. K. and Chuah, C. H. (2011). Characterisation of Galloylated Cyanogenic Glucosides and hydrolysable tannins from leaves of *Phyllagathis rotundifolia* by LC-ESI-MS/MS. *Phytochemical Analysis*, 516–525.

Thakker, V. Y., Shah, V. N., Shah, U. D. and Suthar, M. P. (2011). Simultaneous estimation of Gallic acid, Curcumin and Quercetin by HPTLC method. *Journal of Advanced Pharmacy Education & Research*, 70-80.

Thammana, M. (2016). A Review on High Performance Liquid Chromatography (HPLC). *Journal of Pharmaceutical Analysis*, 22-28.

Theppakorn, T. and Wongsakul, S. (2012). Optimization and Validation of the HPLC-Based Method for the Analysis of Gallic acid, Caffeine and 5 Catechins in Green Tea. *Naresuan University Journal*, 1-11.

Thomas, A., Kanakdhar, A., Shirsat, A., Deshkar, S. and Kothapalli, L. (2020). A high-performance thin layer chromatographic method using a design of experiment approach for estimation of phytochemicals in extracts of *Moringa oleifera* leaves. *Turkish Journal of Pharmaceutical Sciences*, 148–158.

Tiwari, S. and Talreja, S. (2022). Thin Layer Chromatography (TLC) VS. Paper Chromatography : A Review. *Acta Scientific Pharmaceutical Sciences*, 05–09.

Tukiran, T., Mahmudah, F., Hidayati, N. and Shimizu, K. (2016). Gallic acid : A Phenolic acid and its Antioxidant activity from stem bark of Chloroform extracts of *Syzygium Litorale* (blume) Amshoff (Myrtaceae). *Molekul*, 180-189.

Van Bramer, S.E. (1988). An Introduction to Mass Spectrometry.

Vasim, M., Navin, D. and Manish, H. (2016). Simulatneous determination and validation of gallic acid and quercetin in *Anisomeles malabarica* R.Br. Ex Sims using high performance thin layer chromatography. *Journal of Chemical and Pharmaceutical Research* 8, 470-473.

Veer, V., Kalita, P. and Tag, H. (2014). HPTLC determination of gallic acid in methanol extract of Quercus griffithii Acorn. *International Journal of PharmTech Research*, 1341-1347.

Vidushi, Y., Meenakshi, B. and Bharkatiya, M.B. (2017). A review on HPLC Method Developmeent and Validation. *Research Journal of Life Sciences, Bioinformatics, Pharmaceutical and Chemical Sciences*, 166.

Vyas, N. and Patel, S. (2016). Simultaneous Estimation of Curcuminoids, Piperine, and Gallic Acid in an Ayurvedic Formulation by Validated High-Performance Thin Layer Chromatographic Method. *Asian Journal of Pharmaceutical and Clinical Research*, 117–122.

Wang, H., Provan, G. J. and Helliwell, K. (2003). Determination of hamamelitannin, catechins and gallic acid in witch hazel bark, twig and leaf by HPLC. *Journal of Pharmaceutical and Biomedical Analysis*, 539–544.

Wyrepkowski, C. C., da Costa, D. L. M. G., Sinhorin, A. P., Vilegas, W., de Grandis, R. A., Resende, F. A., Varanda, E. A. and dos Santos, L. C. (2014). Characterization and quantification of the compounds of the ethanolic extract from Caesalpinia ferrea stem bark and evaluation of their mutagenic activity. *Molecules*, 16039–16057.

Yisimayili, Z., Abdulla, R., Tian, Q., Wang, Y., Chen, M., Sun, Z., Li, Z., Liu, F., Aisa, H. A. and Huang, C. (2019). A comprehensive study of pomegranate flowers polyphenols and metabolites in rat biological samples by high-performance liquid chromatography quadrupole time-of-flight mass spectrometry. *Journal of Chromatography A*, 1604.

Zakaria, F., Rosli, W. and Ishak, W. (2014). Quantitative HPLC Analysis of Gallic Acid in *Benincasa hispida* Prepared with Different Extraction Techniques. *Sains Malaysiana,* 1181-1187.

Zanwar, A. S., Sen, D. B., Sanathra, R., Dash, K., Maheshwari, R. and Sen, A. K. (2022). Quantitative Assessment of Luliconazole and Gallic Acid Simultaneously in Formulated Emulgel by UV Spectrophotometric Methods, *Asian Journal of Pharmaceutics,* 189-195.

Zhang, L., Bouguet-Bonnet, S. and Buck, M. (2012). Combining NMR and molecular dynamics studies for insights into the allostery of small GTPase-protein interactions. *Methods in Molecular Biology,* 235–259.

Zhao, J., Khan, I. A. and Fronczek, F. R. (2011). Gallic acid. *Acta Crystallographica Section E : Structure,* 316-317.

Chapter 2

A New Perspective on the Efficacy of Gallic Acid in the Treatment of Breast Cancer

Sakshi Patil[*]
Swapnali Patil[†]
Pranali Pangam[‡]
Poournima Sankpal[§]
and Sachinkumar Patil[#]

Ashokrao Mane College of Pharmacy,
Peth-Vadgaon, Kolhapur, Maharashtra, India

Abstract

Gallic acid is (GA) a natural phenolic compound. It is also known as 3,4,5-trihydroxybenzoic acid. It has been suggested that the natural phenolic chemical gallic acid (3,4,5-trihydroxybenzoic acid; GA), which is produced from plants, can stop the growth and spread of a variety of malignancies. Anywhere in the body, cancer is the uncontrolled development of aberrant cells. Malignant, tumor, or cancer cells are all terms used to describe these aberrant cells.

The second biggest cause of mortality in the world is cancer. With more than a million new instances of breast cancer diagnosed worldwide annually, breast cancer is the most prevalent cause of death. Breast cancer is the most common cancer in women that accounts for 33% of all cancers

---

[*] Corresponding Author's Email: 2000sakshipatil@gmail.com.
[†] Corresponding Author's Email: swapnali30may99@gmail.com.
[‡] Corresponding Author's Email: pranalipangam27@gmail.com.
[§] Corresponding Author's Email: poournima6@gmail.com.
[#] Corresponding Author's Email: sachinpatil.krd@gmail.com.

In: The Chemistry of Gallic Acid and Its Role in Health and Disease
Editor: Jeff C. Murdoch
ISBN: 979-8-88697-672-4
© 2023 Nova Science Publishers, Inc.

in women globally. In 2020, 685000 people worldwide died and 2.3 million women were diagnosed with breast cancer. The most common disease in the globe as of the end of 2020 was breast cancer, which had been diagnosed in 7.8 million women in the previous five years.

For breast cancer and melanoma, there are several treatments available, including surgical removal of tumor, radiation therapy, hormone therapy, chemotherapy, targeted biological therapies, etc. Many of them have negative side effects. In order to meet the constant high demand for new anticancer medications, scientists investigate numerous natural and synthetic substances. Gallic acid (GA) found in many dietary substances and herbs used in traditional medicine. It has antibacterial, antiviral, anti-inflammatory, and antioxidant effects. A naturally occurring phenolic compound found in plants called gallic acid (GA) has a number of therapeutic effects that include anti-inflammatory, anti-obesity, and anti-cancer activities.

In more recent studies, gallic acid (GA) has been demonstrated to perform anti-cancer actions through a number of biological mechanisms, including angiogenesis, cell cycle arrest, migration, metastasis, apoptosis, and oncogene expression. In-depth research on gallic acid (GA), anticancer effects in MCF-7 human breast carcinoma cells was done for this review. This article indicated that treatment with gallic acid (GA) resulted in inhibition of proliferation and induction of apoptosis in MCF-7 cells. In conclusion, our review shows that gallic acid (GA) is a unique, powerful, and safe anti-cancer therapeutic candidate for the treatment of breast cancer.

Keywords: gallic acid, breast cancer, MCF-7 cells, cell apoptosis, cell cycle arrest, mitochondrial pathway

Introduction

There are various options for cancer treatment, with surgery, chemotherapy, and radiotherapy being among the top three [1]. The kind, stage, and location of the cancer, the patient's health and preferences, and all of these factors affect the treatment options. Chemotherapy is one of the widely used treatments that employs chemicals to destroy rapidly dividing cells. Chemotherapy is one of the most widely used treatments [2]. Chemotherapeutic drugs make cancer cells go through apoptosis, a sort of programmed cell death that includes biochemical processes that lead to morphological and molecular changes that cause death. It employs chemicals to destroy rapidly dividing cells. Some of the anticancer medications on the market include epirubicin, cisplatin, 5-fluorouracil, doxorubicin, and cyclophosphamide. Even though there are several medications that can delay the development of cancer, when it is discovered in its later stages, it cannot be entirely cured. As a result, the quest for new anticancer medications has been ongoing. In addition to studying synthetic medications, researchers also look into naturally occurring substances found in food [3].

Gallic acid (GA), also known as 3, 4, 5-trihydroxybenzoic acid, C_6H_2 $(OH)_3$ COOH. It is a solid, monomeric phenolic compound that is bonded to sugars and belongs to the hydrolysable tannin family. It has a molecular weight of 170.12 g/mol [4]. Gallic acid (GA) can be found in plant material as free acid, esters, catechin derivatives, and hydrolysable tannins. It is present in relatively high concentrations in a number of vegetal samples and in industrial wastes from where it could be extracted [5]. It can be recovered from industrial wastes and is found in a variety of vegetal samples in relatively high amounts. Additionally, it comes from tannic acid (C76H52O46), which is converted by the extracellular inducible enzyme tannase. The primary enzyme in the oxidation process is tannase, which breaks down the ester linkages to produce gallic acid (GA) esters and hydrolysable tannins [6]. Similar to other natural phenolic compounds, gobernadora and damiana plants, grapes, pomegranates, almonds, strawberries, raspberries, lemons, apple peels, chard, spinach, coffee, wine, and green tea are sources of this one [7]. The extraction of gallic acid (GA) from natural sources has been carried out using polar solvents, such as water, ethanol, and methanol, or mixtures of them [8].

Source and Chemistry of Gallic Acid

Gallic acid (GA) is a trihydroxybenzoic acid, sometimes referred to as 3,4,5-trihydroxybenzoic acid, which is a kind of organic acid [9]. The molecule has the formula C6H2(OH)3COOH. It comes in the form of white, yellowish-white, or light fawn crystals that are soluble in acetone, ether, alcohol, and glycerol. It is an organic acid that is widely known for its high antioxidant capabilities and is present in a broad variety of foods and plants [10]. Numerous foods containing gallic acid (GA) have long been used as natural remedies [11]. Native Americans and early American immigrants, for instance, used blueberries to prepare a fragrant tea that was used as a sedative during delivery and as an excellent tonic for cleaning the blood [12]. Colds, other illnesses, and menstrual irregularities were all treated with the tea alone.

To stop infection, cuts and wounds were treated with hazel balm and tea containing gallic acid (GA) [13]. Chinese herbalists treated intestinal problems, haemorrhage, hematochezia, and hyperhidrosis using gallnuts from oak and sumac. Thirty ayurvedic herbs and formulations have been screened for the presence of gallic acid (GA), which has already been used for years for the treatment of different diseases [14].

By oxidation, gallic acid (GA) is simply released from gallotannins. The fastest way to do it is to use concentrated sulfuric acid to precipitate it out of an aqueous solution. Allowing passive oxidation of ambient oxygen in water is a slower method of generating the gallic acid (GA) [15]. In the pharmaceutical sector, it is mostly utilised for the manufacture of antibacterial medications like trimethoprim. In the food business, gallic acid (GA) is a substrate used in the chemical synthesis of food preservatives such as pyrogallol and gallates [16]. Gallates are the ester derivatives of gallic acid (GA) that are often found in plants, and their biological properties are actively being researched. These investigations showed that gallates promoted apoptosis in many cancer cell types. Alkyl esters were shown to be more effective in inhibiting cancer cell lines than gallic acid (GA) [17]. Lauryl gallate, for instance, was shown to be 40 times more effective than gallic acid (GA) when tested against mouse B-cell lymphoma. The hydrophobic moieties of Wehi 231.30 gallates could be to blame for this result [18]. The drug's affinity for the membrane of cancer cells is increased by the alkyl ester's presence of more than eight carbons [19].

Biological Properties of Gallic Acid

The antimalarial action of gallic acid (GA) was the first medical property to be described [20]. Later, it was shown that gallic acid (GA) had antifungal properties. In the same year, researchers looked at the antibacterial efficacy of synthetic gallic acid (GA) derivatives against *Escherichia coli*, *Staphylococcus aureus*, and *Bacillus subtilis* [21]. Gallic acid (GA's) antiviral activity was demonstrated by *in vivo* and *in vitro* studies on the mortality of monkeys exposed to influenza. The many and important roles of gallic acid (GA) were established thanks to all of these important findings, giving this important natural molecule a new situation [22]. A rise in DNA damage and the release of cytochrome c are signs that gallic acid (GA) has potent antioxidant effects. It also decreased mitochondrial potential and glutathione levels in the cells [23]. It also has a dose-dependent antifungal action [24]. In cases of internal bleeding, as well as to treat albuminuria, diabetes, and other disorders, it is used as a remote astringent. Gallic acid (GA) showed cytotoxicity toward cancer cells without affecting healthy cells, researchers found. Epigallo catechin gallate (EGCG), lauryl gallate, and theaflavin-3-gallate stand out among the several alkyl derivatives of gallic acid (GA) that also have anticancer properties [25]. This article's major focus is on the anticancer capabilities of gallic acid (GA) and its derivatives against various cancer cell lines [26]. Gallic acid (GA) also works effectively to reduce inflammation. Gallic acid (GA), according to studies, inhibited both the production of inflammatory indicators and the activation of NF-B dependent p65 acetylation. Due to the complete lack of NF-B activity brought on by the low rate of p65 acetylation, gallic acid (GA) was developed into a unique anti-inflammatory drug [27].

Bioavailability and Toxicity of Gallic Acid

Before a medicine is added to chemotherapy, it should be studied for consequences outside of its anticancer activity. Two crucial ones are toxicity and bioavailability [28]. Bioavailability is the study of a given compound's bioavailability for physiological function, frequently in an *in vivo* environment. Identification of the compound's metabolites after absorption is made easier by this research. Bioavailability and toxicity of the molecule must be investigated prior to standardizing a chemotherapeutic drug. The bioavailability of gallic acid (GA) has been investigated in both animal models and human studies [29]. 4-O-methyl gallic acid (4OMGA), which results from O-methylation, is the major metabolite seen in the urine of rats or rabbits eating gallic acid (GA), propyl gallate, lauryl gallate, or tannic acid. Rats were given gallic acid (GA) orally, and the intestinal absorption of gallic acid (GA) was examined. The rats were given 100 μmol L−1 body weight of gallic acid (GA). Gallic acid (GA) was slowly absorbed and 0.71 μmol L−1 of GA and its metabolite 4OMGA was found in serum. Gallic acid (GA) and its metabolites may be found in human plasma and urine by taking tablets containing 50 mg of acidum gallicum [30].

The experiment's results showed that 4OMGA and unmodified gallic acid (GA) may be found in body fluids, including plasma and urine. Following that, a different study was carried out to evaluate the pharmacokinetics and bioavailability of gallic acid (GA) in healthy people. The individuals either received acidum gallicum tablets (10% gallic acid (GA) and 90% glucose) or black brew tea (0.3 mmol GA). Compared to 39.6-5.1% for black tea, 36.4-4.5% of unmodified gallic acid (GA) and its metabolite 4OMGA were found in the urine after consuming acidum gallicum tablets [31]. It was determined that the bioavailability of gallic acid (GA) was 1.06-0.26 when comparing taking pills and drinking tea, proving that it was not affected by the matrix of distribution [32]. This article claims that gallic acid (GA) may be administered orally throughout chemotherapy in the form of a pill or in free form.

The bioavailability and efficacy of antioxidants such as gallic acid (GA), quercetin, epigallocatechin gallate (EGCG), and n-propyl gallate in human corneal limbal epithelial (HCLE) cells were investigated to see if they would be valuable components of lubricating eye drops. The production of ROS was significantly reduced when an antioxidant was present both in the media with the xanthine oxidase and within the cells. They could be helpful in reducing oxidative damage to the corneal epithelium since they are bioavailable, according to this theory. When taking a medication, it's also important to consider toxicity, which is the word used to describe how a drug affects the body as a whole [33].

The toxicity of gallic acid (GA) was investigated using the mouse model, and a NOAEL (no observable adverse effect threshold) was also determined. The NOAEL was determined to be the maximal dosage of 5000 mg kg^{-1} taken orally, which showed no significant changes in the haematological parameters. By feeding F344 rats diets containing 0, 0.2, 0.6, 1.7, and 5% gallic acid (GA) for 13 weeks, subchronic toxicity of gallic acid (GA) was examined [34]. Clinical symptoms, body weight, food intake, haematology, blood biochemistry, organ weights, and histological evaluation were all considered toxicological markers. From the first week of the trial to the end, both sexes of the animals receiving 5% gallic acid (GA) gained weight. Males and females who received 0.6% or more experienced toxic effects, which decreased haemoglobin concentration, hematocrit, and red blood cell counts and increased reticulocytes [35]. Histopathological analysis revealed the development of hemolytic anaemia. Additionally, 1.7% more liver weight was measured due to centrilobular liver cell hypertrophy.

These toxicological findings led to the conclusion that 0.2% was a NOAEL in rats. For male and female rats, this amount corresponded to 119 and 128 mg per kg per day, respectively. Similar to this, much research has been done on the toxicity of GA's propyl, octyl, and dodecyl esters in animal models that involve oral administration. This study demonstrated that propyl gallate's biokinetics differed from those of octyl and dodecyl gallate due to its degree of hydrolysis and absorption. The induction of liver enzymes was seen at 5000 mg of propyl gallate per kilogramme of feed. In contrast, the dosage of 3000 mg kg1 meal or greater was required for the octyl gallate or dodecyl gallate to exhibit effects. In conclusion, the FAO/WHO Joint Expert Committee on Food Additives (JECFA) approved 0.2 mg kg1 body weight (as a total of propyl, octyl, and dodecyl gallates) as an acceptable daily intake (ADI) for males and 1000 mg kg1 feed as a no-effect threshold [36].

In 2022, women will be diagnosed with 287,850 new instances of invasive breast cancer, 51,400 new cases of DCIS, and 43,250 new cases of breast cancer-related deaths. Women over the age of 50 account for 83% of invasive breast cancer diagnoses, and same age group accounts for 91% of breast cancer death. Half of breast cancer deaths occur in women 70 years or old. The median age at diagnosis for female breast cancer is 62 years overall but is slightly younger for Hispanic (57 years), API (58 years), Black (60 years), and AIAN (61 years) women than for White women (64 years) 6, due in part to variations in the population's age distribution. The average age at breast cancer mortality is 69 years, however it is 70 years for White women, 62 years for Hispanic women, 63 years for API and Black women, and 62 years for Black women. Although breast cancer is primarily a female disease, approximately 2710 cases and 530 deaths (approximately 1% of all breast cancer cases and deaths) are expected in men in 2022. The information provided herein applies to female breast cancer unless otherwise specified [37].

The breast is made up of a variety of tissues, from extremely fatty tissue to extremely dense tissue. An intricate lobe network exists inside this tissue. Each lobe is composed of tiny, tube-like structures called lobules that house milk glands. Milk is transported from the lobes to the nipple via small ducts that link the glands, lobules, and lobes. The areola, the darker region surrounding the nipple, contains the nipple in the center. Additionally, the breast is covered in lymphatic and blood vessels. By bringing oxygen and nutrition to the cells as well as eliminating waste and carbon dioxide, blood vessels nurture the cells. Unlike blood vessels, lymph vessels exclusively transport fluid away from tissues. They join the lymphatic system, which removes body waste, and the lymph nodes. The tiny, bean-shaped structures known as lymph nodes aid in the defence against infection [38]. The body has several locations for groups of lymph nodes, including the neck, groyne, and belly. Regional lymph nodes of the breast are those that are close to the breast, such as the axillary lymph nodes beneath the arm. Cancer begins when healthy cells in the breast change and grow out of control, forming a mass or sheet of cells called a tumor. A tumour can be cancerous or benign. A cancerous tumour is malignant, meaning it can grow and spread to other parts of the body. A benign tumour means the tumour can grow but has not spread [39].

Stages I, II, and III of invasive breast cancer, which includes non-invasive (stage 0) and early-stage cases, are all covered in this handbook. The stage of breast cancer indicates how far the disease has progressed and whether or not it has spread. Breast cancer can spread through the blood vessels and/or lymph nodes to places including the bones, lungs, liver, and brain; however, it most frequently travels to surrounding lymph nodes, in which case it is still regarded as a local or regional illness. This is the most advanced stage of the illness and is known as metastatic or stage IV breast cancer. The presence of adjacent lymph nodes alone, however, does not often indicate stage IV breast cancer [40].

Signs and Symptoms of Breast Cancer [41-47]

In its early stages, breast cancer may not exhibit any symptoms. In many cases, a tumor may be too small to be felt, but an abnormality can still be seen on a mammogram. The initial indication of a tumour, if one can be felt, is typically a new lump in the breast that wasn't there before. But not every lump is cancer. A number of symptoms can be brought on by various breast cancer types. While many of these symptoms are comparable, others can differ. The following are signs of the most prevalent breast cancers:

- A new, distinct-feeling breast lump or tissue thickening that feels different from the surrounding tissue.
- Breast pain
- Red or discolored, pitted skin on the breast
- Swelling in all or part of your breast
- A nipple discharge other than breast milk
- Bloody discharge from your nipple
- Peeling, scaling, or flaking of skin on your nipple or breast
- A sudden, unexplained change in the shape or size of your breast
- Inverted nipple
- Changes to the appearance of the skin on your breasts
- A lump or swelling under your arm

Types of Breast Cancer [41-47]

Breast cancer can be invasive or non-invasive. Invasive breast cancer is cancer that spreads into surrounding tissues and/or distant organs. Non-invasive breast cancer does not go beyond the milk ducts or lobules in the breast. Breast cancers come in a variety of forms and are classified according to how they appear under a microscope.

- *Ductal carcinoma:* This is the most common type of breast cancer.
- *Ductal carcinoma in situ (DCIS):* This is a non-invasive cancer (stage 0) that is located only in the duct and has not spread outside the duct.
- *Invasive or infiltrating ductal carcinoma:* This is cancer that has spread outside of the ducts or lobules.
- *Invasive lobular carcinoma:* This is a less commonly occurring type of breast cancer that has spread outside of the ducts or lobules.

Less common types of breast cancer include:

- Medullary
- Mucinous
- Tubular
- Metaplastic
- Papillary
- Micropapillary
- Apocrine
- Inflammatory breast cancer, which is an aggressive type of cancer that accounts for about 1% to 5% of all breast cancers.
- Paget's disease, which is a rare type of cancer in the skin of the nipple or in the skin closely surrounding the nipple. It begins in the ducts of the nipple, then spreads to the nipple surface and the areola (dark circle of skin around the nipple). The nipple and areola often become scaly, red, itchy, and irritated. Although it is usually non-invasive, it can also be an invasive cancer. It is usually found with an underlying breast cancer.

Therapeutic Role of Gallic Acid

The majority of malignancies in women, or 33% of all cancers in women worldwide, are breast cancers. Breast cancer treatment frequently involves the use of many therapeutic modalities, including local therapy (such as surgery and radiation), systemic therapy (such as chemotherapy, hormone therapy, and biologic or targeted treatments), or a combination of both [45, 46]. Breast cancer is a biologically varied and heterogeneous illness. Treatment is effective for a wide range of illnesses. However, 15% of all breast cancers that do not respond well to treatment are triple-negative (TNBC), and metastasis is a major cause of TNBC cancer mortality [47, 48, 49].

Gallic acid (GA) had a cytotoxic effect on human breast cancer MCF-7 cells, but this effect was probably mitigated by inducing cell death. They looked at how the intrinsic and extrinsic signaling pathways were associated with apoptosis in MCF-7 cells in order to gain insight into the molecular mechanism of gallic acid (GA's) effect on cell death. They showed that after gallic acid (GA) treatment, MCF-7 cells had increased Fas/Fas ligand (FasL) protein levels and initiator caspase-8 activity. Thus, it was demonstrated that the Fas/FasL apoptotic framework was interested in GA-induced cell death. They discovered that gallic acid (GA) therapy resulted in mitochondrial dysfunction by decreasing mitochondrial membrane potential, increasing Bax/Bcl-2 (B cell lymphoma 2) percentage, cytochrome c release into cytosol, and also activating caspase-9 [50].

Further study demonstrated that Ziyang green tea (ZTP), which contains (-) -epigallocatechin gallate (28.2%), (-) -epigallocatechin (5.7%), and (-) -epicatechingallate (12.6%), inhibits the growth of MCF-7 cells by preventing cell cycle progression in the G0/G1 stage and inducing apoptosis. In MCF-7 cells, ZTP encourages cell-cycle arrest by upregulating p53 and downregulating CDK2. The overproduction of reactive oxygen species (ROS) caused by ZTP treatment of MCF-7 cells suggests that ROS play a crucial role in the activation of apoptosis in MCF-7 cells [51]. Further study demonstrated that EGF-treated cells' migration and invasion were markedly inhibited by GA-capped gold nanoparticles (GA-AuNPs), which also prevented the up-regulation of the EGF-induced matrix metalloproteinase-9 (MMP-9). In EGF-treated breast cancer cells, GA-AuNPs suppress EGF-dependent Akt/p65 and ERK/c-Jun phosphorylation, leading to down-regulation of MMP-9 mRNA and protein expression. Therefore, it is assumed that GAAuNPs have a greater capacity than gallic acid (GA) to suppress the production of MMP-9 mediated by EGF/EGFR in TNBC MDA-MB-231 cells [52]. By inhibiting the

PI3K/AKT pathway and regulating the expression of miR-126, mango polyphenolics, which contain gallic acid (GA) as a bioactive component, exhibit anti-carcinogenic activities in BT474 breast cancer cells both *in vitro* and in xenografts [53]. The MCF-7 human breast cancer cell's cell proliferation is largely reduced by gallic acid (GA) treatment in a measurement-dependent manner. It causes a noticeable G2/M phase arrest but very slightly affects the population of sub-G1 MCF-7 cells. The disruption of the p27Kip1/Skp2 pathway and the subsequent proteosomal degradation of p27Kip1 may also increase the amount of p27Kip1, causing the arrest of the MCF-7 cell in the G2/M phase. Gallic acid (GA) is advised to be advantageous in the treatment of breast and p27Kip1-deficient cancer [50, 54]. In general, this research suggests that gallic acid (GA) is capable of treating breast tumours as a restorative specialist.

Conclusion

Gallic acid (GA) is a powerful phenolic molecule having pharmacological properties. Based on newly published studies, the anti-tumor action of gallic acid (GA) is highlighted in this article. With prospective uses in combination chemotherapy and immunotherapy, gallic acid (GA) has demonstrated itself as a potential agent in reducing the proliferation and spread of cancer cells. In cancer treatment, gallic acid (GA) induces apoptosis, ferroptosis, and necroptosis. Gallic acid (GA) induces lipid peroxidation and necroptosis, which lead to ferroptosis.

References

[1] Rampling R, James A, and Papanastassiou V, *J. Neurol., Neurosurg. Psychiatry*, 2004, 75, 24–30.
[2] Lind MJ. *Medicine*, 2008, 36(1), 19–23.
[3] Joensuu H., *Lancet Oncol.*, 2008, 9(3), 304.
[4] Govea-Salas M, González-Castillo MAM, Aguilar-González CN, Rodríguez-Herrera R, Zugasti-Cruz A, Silva-Belmares SY, Rodríguez-Herrera R, Belmares-Cerda R, Morlett-Chávez J. Gallic acid extraction and its application to prevention and treatment of cancer. *Handbook of Gallic Acid*. Nova Science Publishers; 2013 p. 157–177.

[5] Cláudio AFM, Ferreira AM, Freire CSR, Silvestre AJD, Freire MG, Coutinho JAP. Optimization of the gallic acid extraction using ionic-liquid-based aqueous two-phase systems. *Separation and Purification Technology* 2012;97:142–9.
[6] Bravo L. Polyphenols: chemistry, dietary sources, metabolism, and nutritional significance. *Nutr Rev* 1998;56:317–33.
[7] Taitzoglou IA, Tsantarliotou M, Zervos I, Kouretas D, Kokolis NA. Inhibition of human and ovine acrosomal enzymes by tannic acid *in vitro. Reproduction* 2001;121:131–7.
[8] Rakesh SU, Salunkhe VR, Dhabale PN, Burade KB. HPTLC Method for quantitative determination of gallic acid in hydroalcoholic extract of dried f lowers of Nymphaea Stellata Willd. *Asian J Res Chem* 2009;2(2):131–4.
[9] Ayaz FA, Hayirlioglu AS, Gruz J, Novak O and Strnad M, *J. Agric. Food Chem.*, 2005, 53(21), 8116–8122.
[10] Wang H, Provan GJ and Helliwell K, *J. Pharm. Biomed. Anal.*, 2003, 33(4), 539–544.
[11] Lua, JJ, Weib Y.and Yuan QP, *Sep. Purif. Technol.*, 2007, 55, 40–43.
[12] Borde VU, Pangrikar PP and Tekale SU, *Recent Res. Sci. Technol.*, 2011, 3, 51–54.
[13] Locatelli C, Filippin-Monteiro FB and Creczynski-Pasa TB, *Eur. J. Med. Chem.*, 2013, 60, 233–239.
[14] Mukherjee G and Banerjee R, *Chem. Today*, 2003, 21, 59–62.
[15] Ohno Y, Fukuda K, Takemura G, Toyota M, Watanabe M, Yasuda N, Xinbin Q, Maruyama R, Akao S, Gotou K, Fujiwara T and Fujiwara H., *Anticancer Drugs*, 1999, 10(9), 845–851.
[16] Kim NS, Jeong SI, Hwang BS, Lee YE, Kang SH, Lee HC and Oh CH, *J. Med. Food*, 2011, 14(3), 240–246.
[17] Serrano C, Palacios G, Roy C, Cespón ML, Nocito VM and González PP, *Arch. Biochem. Biophys.*, 1998, 350(1), 49–54.
[18] Inoue M, Sakaguchi N, Isuzugawa K, Tani H and. Ogihara Y, *Biol. Pharm. Bull.*, 2000, 23(10), 1153–1157.
[19] Singh M, Bhui K, Singh R and Shukla Y, *Life Sci.*, 2013, 93(1), 7–16.
[20] Thompson PE, Moore AM and Reinertson JW, *Antimicrob. Agents Chemother.*, 1953, 3, 399–408.
[21] MahadevanA, and Reddy MK, Neth. *J. Plant Pathol.*, 1968, 74, 87–90.
[22] Florov F and Mishenkova EL, *Mikrobiol. Zh.*, 1970, 32, 628–633.
[23] Claudriana B, Filippin FM and Ariana C, Antioxidant, antitumoral and anti-inflammatory activities of Gallic acid, Handbook on Gallic Acid: Natural Occurrences, Antioxidant Properties and Health Implications, Chapter: *Antioxidant, Antitumoral and Anti-Inflammatory Activities of Gallic Acid*, 4th edn, 2013, pp. 1–23.
[24] Nguyen DM, Seo DJ, Lee HB, Kim IS, Kim KY, Park RD and Jung WJ, *Microb. Pathog.*, 2013, 56, 8–15.
[25] van der Heijden A, Janssen P, and Strik JJ, *Food Chem. Toxicol.*, 1986, 24, 1067–1070.
[26] Fiuza SM, Gomes C, Teixeira LJ, Girão da Cruz MT, Cordeiro MNDS, Milhazes N, Borges F, Marques MPM, Choi KC, Lee YH, Jung MG, Kwon SH, Kim J, Jun WJ,

and Lee J, Phenolic acid derivatives with potential anticancer properties—a structure–activity relationship study. Part 1: Methyl, propyl and octyl esters of caffeic and gallic acids, *Mol. Cancer Res.*, 2009, 7, 2011–2021.

[27] Choi KC, Lee YH, Jung MG, Kwon SH, Kim J, Jun WJ, and Lee J, *Mol. Cancer Res.*, 2009, 7, 2011–2021.

[28] Kaliora AC, Kanellos PT, and Kalogeropouls N, Chapter: Gallic acid bioavailability in humans, in *Handbook on Gallic Acid*, Nova Publishers, 2013.

[29] Robbins FD, Emerson OH, Jones FT, Booth AN and Masri MS, *J. Biol. Chem.*, 1959, 234, 3014–3016.

[30] Konishi Y, Hitomi Y, and Yoshika E., *J. Agric. Food Chem.*, 2004, 52, 2527–2532.

[31] Shahrzad S, and Bitsch I, Chromatogr J. B: *Biomed. Sci. Appl.*, 1998, 705, 87–95.

[32] Shahrzad S, Ayogi K, Winter A, Koyama A. and Bitsch I., *J. Nutr.*, 2001, 131(4), 1207–1210.

[33] Stoddard R, Koetje LR, Mitchell AK, Schotanus MP, and Ubels JL, *J. Ocul Pharmacol. Ther.*, 2013, 29(7), 681–687.

[34] Rajalakshmi K, Devaraj H, and Niranjali Devaraj S, *Food Chem. Toxicol.*, 2001, 39, 919–922.

[35] Niho N, Shibutani M, Tamura T, Toyoda K, Uneyama C, Takahashi N and Hirose M, *Food Chem. Toxicol.*, 2001, 39, 1063–1070.

[36] van der Heijden A, Janssen PJ, and Strik JJ, *Food Chem. Toxicol.*, 1986, 24(10–11), 1067–1070.

[37] Surveillance, Epidemiology, and End Results (SEER) Program. SEER*Stat Database: Incidence—SEER Research Data, 17 Registries, November 2021 Submission (2000–2019)—Linked To County Attributes—Time Dependent (1990–2019) Income/Rurality, 1969–2020 Counties. Surveillance Research Program, Division of Cancer Control and Population Sciences, National Cancer Institute; 2022. Accessed July 1, 2022.

[38] Surveillance, Epidemiology, and End Results (SEER) Program. SEER*Stat Database: Mortality—All Causes of Death, Aggregated With State, Total U.S. (1990–2020) <Katrina/Rita Population Adjustment>. Surveillance Research Program, Division of Cancer Control and Population Sciences, National Cancer Institute; 2022. Underlying mortality data provided by the National Center for Health Statistics. Accessed July 1, 2022.

[39] Siegel RL, Miller KD, Fuchs HE, Jemal A. Cancer statistics, 2022. CA *Cancer J Clin*. 2022;72:7-33.

[40] Yadav S, Karam D, Bin Riaz I, Xie H, Durani U., Duma N, Giridhar KV, Hieken TJ, Boughey JC, Mutter RW, Hawse JR, Jimenez RE, Couch FJ, Leon-Ferre RA, & Ruddy KJ. Male breast cancer in the United States: treatment patterns and prognostic factors in the 21st century. *Cancer*. 2020;126:26-36.

[41] Guray M, Sahin AA. Benign breast diseases: classification, diagnosis, and management. *Oncologist*. 2006 May;11(5):435-49.

[42] Wysokinska EM, Keeney G. Breast cancer occurring in the chest wall: rare presentation of ectopic milk line breast cancer. *J Clin Oncol*. 2014 Apr 01;32(10):e35-6.

[43] Ju DG, Yurter A, Gokaslan ZL, Sciubba DM. Diagnosis and surgical management of breast cancer metastatic to the spine. *World J Clin Oncol.* 2014 Aug 10;5(3):263-71.

[44] Al-Gaithy ZK, Yaghmoor BE, Koumu MI, Alshehri KA, Saqah AA, Alshehri HZ. Trends of mastectomy and breast-conserving surgery and related factors in female breast cancer patients treated at King Abdulaziz University Hospital, Jeddah, Saudi Arabia, 2009-2017: A retrospective cohort study. *Ann Med Surg* (Lond). 2019 May;41:47-52.

[45] Sarhadi NS, Shaw Dunn J, Lee FD, Soutar DS. An anatomical study of the nerve supply of the breast, including the nipple and areola. *Br J Plast Surg.* 1996 Apr;49(3):156-64.

[46] Sarhadi NS, Shaw-Dunn J, Soutar DS. Nerve supply of the breast with special reference to the nipple and areola: Sir Astley Cooper revisited. *Clin Anat.* 1997;10(4):283-8.

[47] Schulz S, Zeiderman MR, Gunn JS, Riccio CA, Chowdhry S, Brooks R, Choo JH, Wilhelmi BJ. Safe Plastic Surgery of the Breast II: Saving Nipple Sensation. *Eplasty.* 2017;17:e33.

[48] Network TCGA. Comprehensive molecular portraits of human breast tumours. *Nature.* 2012;490:61.

[49] Anjum F, Razvi N, Masood MA. Breast cancer therapy: a mini review. *MOJ Drug Des Dev Ther.* 2017;1:00006.

[50] Lee JC, Chen WC, Wu SF, Tseng Ck, Chiou CY, Chang FR, Hsu S, & Wu C. (2011) Anti-hepatitis C virus activity of Acacia confusa extract via suppressing cyclooxygenase-2. *Antiviral Res* 89: 35-42.

[51] Wang C, Kar S, Lai X, Cai W, Arfuso F, Sethi G, Lobie PE, Goh BC, Lim LHK, Hartman M, Chan CW, Lee SC, Tan SH, & Kumar AP. Triple negative breast cancer in Asia: an insider's view. *Cancer Treat Rev.* 2018;62:29–38.

[52] Kang N, Lee JH, Lee W, Ko JY, Kim EA, Kim JS, Heu MS, Kim GH, & Jeon YJ. (2015) Gallic acid isolated from Spirogyra sp. improves cardiovascular disease through a vasorelaxant and antihypertensive effect. *Environ Toxicol Pharmacol* 39: 764-772.

[53] Kyriakis E, Stravodimos GA, Kantsadi AL, Chatzileontiadou DS, Skamnaki VT, & Leonidas DD. (2015) Natural flavonoids as antidiabetic agents. The binding of gallic and ellagic acids to glycogen phosphorylase b. *FEBS Lett* 589: 1787-1794.

[54] Latha RCR, Daisy P (2011) Insulin-secretagogue, antihyperlipidemic and other protective effects of gallic acid isolated from Terminalia bellerica Roxb. in streptozotocin-induced diabetic rats. *Chem Biol Interact* 189: 112-118.

Chapter 3

A New Perspective on the Efficacy of Gallic Acid in the Treatment of Lung Cancer

Swapnali Patil[*]
Pranali Pangam[†]
Shravan Joshi[‡]
Poournima Sankpal[§]
and Sachinkumar Patil[ǁ]
Ashokrao Mane College of Pharmacy,
Peth-Vadgaon, Kolhapur, Maharashtra, India

Abstract

According to estimates from the International Agency for Research on Cancer, 1 in 5 individuals worldwide will acquire cancer at some point in their lifetime. As per United States study reports, it is estimated that there will be 1,918,030 new cancer cases and 609,360 cancer deaths in 2022, with lung cancer being the primary cause of death accounting for around 350 of those fatalities daily. Clinical interventions have had very little chance of reducing lung cancer-related deaths in recent years.

More than 80% of occurrences of lung cancer are non-small-cell lung cancer (NSCLC), which is the most prominent subtype, whereas small cell lung cancer (SCLC) makes up more than 15% of all occurrences of lung cancer and is the most dangerous subtype of the

---

[*] Corresponding Author's Email: swapnali30may99@gmail.com.
[†] Corresponding Author's Email: pranalipangam27@gmail.com.
[‡] Corresponding Author's Email: shravanjoshi1999@gmail.com.
[§] Corresponding Author's Email: poournima6@gmail.com.
[ǁ] Corresponding Author's Email: sachinpatil.krd@gmail.com.

In: The Chemistry of Gallic Acid and Its Role in Health and Disease
Editor: Jeff C. Murdoch
ISBN: 979-8-88697-672-4
© 2023 Nova Science Publishers, Inc.

disease. A variety of molecular mechanisms and biological pathways play an important role in the development of lung cancer therapies. In order to prevent and treat non-small-cell lung cancer (NSCLC) and small cell lung cancer (SCLC), polyphenols are the naturally occurring sources of potential cancer chemotherapeutic agents that can minimize the side effects of conventional anticancer medications and aid in the fight against drug resistance.

Gallic acid (GA) is gaining significant interest as a form of phenolic acid due to its excellent bioavailability and non-toxicity. The most recent research on gallic acid's anti-tumor properties in various malignancies was examined, with an emphasis on the molecular mechanisms and cellular pathways that lead to tumor cell apoptosis and migration. When gallic acid and chemotherapeutic drugs are administered simultaneously, tumor proliferation is suppressed more effectively. In this chapter, we extensively studied the anticancer effect of gallic acid in human SCLC H446 cell line and NSCLC EGFR-TKI (tyrosine kinase inhibitor)-resistant cell line. This article indicated that treatment with gallic acid (GA) resulted in inhibition of proliferation and induction of apoptosis in NSCLC and SCLC cells. According to this investigation, evidence suggests that gallic acid is a promising new, potent, and safe anti-cancer therapeutic candidate for the treatment of lung cancer.

Keywords: gallic acid, lung cancer, NSCLC, SCLC, polyhydroxy phenolic substance

Introduction

Cancer is a non-communicable illness that ranks second in the world behind cardiovascular diseases as a cause of mortality. Cancer is the main obstacle to extending life expectancy since its occurrence is closely tied to ageing (Abdel-Razeq, Hikmat, et al. 2015). Healthcare practitioners in economically developed and developing nations equally face significant challenges in the management and treatment of cancer. Planning, targeted therapies, and timelines must be carefully considered while treating cancer.

Cancer still has a high mortality and morbidity rates in the United States despite a significant rise in the number of survivors (Miller, Kimberly D., et al. 2022). Furthermore, some lifestyle modifications, such as quitting smoking, have decreased the incidence of cancer-related fatalities. Different genetics, growth and migratory patterns, sensitivity to therapy, and other characteristics set cancer cells apart from normal cells. Each type of carcinoma has distinct standard treatment choices, although surgery, chemotherapy, and radiation are the most often used regimens (Golemis, Erica A, Paul Scheet, et al. 2018). The effectiveness of surgery is frequently limited by the spread of tumor cells into nearby and organs and distant tissues. Surgery cannot completely eradicate all cancer cells because they can spread to different places of the body, especially at more advanced stages. Additionally, surgery is an intrusive procedure; moreover, its use is not always advised. On the contrary, chemotherapy and radiotherapy are inefficient against cancer cells that have advanced in their proliferation. The principal objectives in the field of cancer research are the development of innovative anti-tumor drugs that are very effective in reducing cancer progression and proliferation with minimal toxicity. There are adverse effects, therapeutic resistance, and high costs associated with the present arsenal of synthetic medicines with anti-tumor properties (Ashrafizadeh, Milad, Ali Zarrabi, et al. 2021).

The majority of current research focused on the causes and potential remedies for cancer drug resistance. Accordingly, cancer cells that are resistant to chemotherapy require an urgent need for innovative anti-cancer substances. Phytochemicals have significant importance in development of cancer therapy. Natural compounds made from plants have a low bioavailability, which limits their anti-tumor efficacy. Gallic acid (GA) is a phenolic acid that can only be obtained naturally from plants like gallnut, sumac, tea, and oak bark (Subramanian, A. P., A. A. John, et al. 2015).

Lung Cancer

Lung cancer is the leading cause of cancer-related morbidity and mortality globally, with an estimated two-million cases diagnosed and 1.8 million deaths per year (Štěpánek, Ladislav, Jarmila Ševčíková, et al. 2022). Lung cancer is the second most frequent cancer diagnosed in both men and women. Globally, the incidence of lung cancer is rising due to expanding tobacco access and industrialization in emerging countries (Barta, Julie A., Charles A. Powell, et al. 2019). The typical diagnostic age is 70 years old. Males are more likely than females to develop lung cancer due to tobacco use, but women may be more susceptible due to higher frequencies of mutations in the epidermal growth factor receptor as well as the influence of oestrogen (Thandra, Krishna Chaitanya, Adam Barsouk, et al. 2021). Males of African American heritage are most at risk for lung cancer in the United States. When there is a genetic predisposition, the risk is 1.7 times more likely, with first-degree relatives having a greater risk.

Up to 90% of occurrences of lung cancer are attributable to tobacco use, and prolonged usage is expected to increase the prevalence of cancer globally, especially in developing countries like China, Russia, and India. Children and spouses have both been linked to secondhand smoking. The second most common cause of lung cancer in the industrialized world is radon from naturally occurring underground uranium decay (Xu, Xing, Mahdi Fallah, et al. 2020). Smoking marijuana, using electronic cigarettes, using heated tobacco products, and being infected with HIV, COVID-19 and TB have all been linked to an increased risk of lung cancer, as have occupational risks like asbestos and environmental exposures like air pollution and arsenic (Jain, Anuj, Gaurav Phull, et al. 2021).

The most prevalent and fatal types of cancer globally are lung neoplasms (Fitzmaurice, Christina, Christine Allen, et al. 2017), Small-cell lung cancer (SCLC) and non-small-cell lung cancers (NSCLC), are the two categories of lung cancer depending on the cell of origin. The 2015 WHO classification lists adenocarcinoma (cancer of glandular cells), squamous cell carcinoma (SCC), and neuroendocrine malignancies such small cell carcinoma (SCLC), large cell neuroendocrine carcinoma (LCNEC), and carcinoid as the most prevalent kinds of lung cancer. Since SCLC also develops from poorly differentiated neuroendocrine cells, it has a poor prognosis, rapid metastasis, and poor therapeutic response.

Carcinoid tumors are malignancies of well-differentiated neuroendocrine cells (Kulchitsky cells) (Travis, William D., Elisabeth Brambilla, et al. 2015). Squamous cell and small cell malignancies, especially in men, are more likely to be centrally situated and linked to a history of smoking. Women and those without a history of smoking are more likely to develop adenocarcinoma, which is characterized by peripheral onset and the presence of targetable driver mutations in the EGFR, ALK, BRAF, and ROS1 genes.

Recent years have seen the replacement or augmentation of chemotherapy in eligible patients by receptor tyrosine kinase small molecule inhibitors against these mutations and immunotherapies like programmed cell death protein 1 (PD-1) and cytotoxic T-lymphocyte-associated protein 4 (CTLA-4) inhibitors (Denisenko, Tatiana V., Inna N. Budkevich, 2018).

Smoking is to blame for more than 80% of lung cancer occurrences in the west, and success in quitting smoking has led to decreases in incidence and death (Gredner, Thomas, Ute Mons, 2021). In the developing world, smoking still contributes significantly, along with additional risk factors such asbestos and combustion gases exposure at work, as well as arsenic exposure from the environment and air pollution. A better understanding of the epidemiology and risk factors for lung cancer can guide preventative strategies and reduce the rising disease burden globally (Hashim, Dana, and Paolo Boffetta, 2014).

Gallic Acid

Polyphenols are a structurally varied category of compounds that are abundantly present in nature, particularly in plants. The most prevalent polyphenols, condensed tannins, are found in almost all plant groups, even lower plants like mushrooms (Schultz, Jack C., Mark D. Hunter, et al. 1992). Some polyphenols, like phytoalexins produced by plants in response to pathogen infections, exhibit hypersensitive reaction (Gurib-Fakim and Ameenah, 2006). In areas close to processing industries for the coffee, paper, carpet, and olive oil industries, polyphenols can be a source of pollution. On the other hand, a number of polyphenols are well recognized for their antioxidant, anti-inflammatory, and anticancer properties, which warrant consideration in research on the manufacturing of cancer drugs.

Gallic acid (3,4,5-trihydroxicbenzoic acid), a well-known polyphenol, is significant in this regard since it is specifically cytotoxic to a variety of tumor cells (Verma, Sharad, Amit Singh, et al. 2013). Gallic acid is an abundant endogenous plant polyphenol that may be found in wine, tea, grapes, berries, and other fruits (Brglez Mojzer, Eva, Maša Knez Hrnčič, et al. 2016). Some hard wood plant species, including oak, chestnut, and many others, also contain gallic acid (Patil, P. and S. Killedar, et al., 2021). It is a white or yellowish crystal with a melting point of 250°C and a 20°C water solubility of 1.1%.

Gallic acid has potent antioxidant, anti-inflammatory, antimutagenic, and anticancer activities and is known to alter a number of pharmacological and biochemical processes (Patil, P. and S. Killedar, et al. 2022). In the presence of metal ions, gallic acid is also known to have pro-oxidant properties that vary on concentration. Gallic acid's pro-oxidant qualities have been identified as an apoptosis inducer in cancer cell types. The preventive effect of gallic acid in chemically induced carcinogenesis has also been confirmed by studies (Choubey, Sneha, Lesley Rachel Varughese, et al. 2015). At different molecular levels, gallic acid affects different tumor types in a variety of ways. This compound is an essential biomolecule for therapeutic usage due to its effects on cancer (Verma, Sharad, Amit Singh, et al. 2013).

Gallic acid has relatively minimal toxicity for normal cells and demonstrated specific cytotoxicity for malignant cells. Gallic acid is a significant dietary supplement and vitamin supplement because of its feature, which lowers the chance of developing cancer. This study explores new avenues in anticancer research while methodically highlighting the benefits of gallic acid suppressing malignant cells and the multi-targeted molecular pathways explaining chemopreventive and therapeutic characteristics (Choubey, Sneha, Soniya Goyal, et al. 2018).

Chemistry of Gallic Acid

Gallic acid is a naturally occurring polyphenolic molecule that is processed and found in red wines and green teas (3, 4, 5-trihydroxybenzoic acid). In plants, it manifests as unbound acids, esters, catechin derivatives, unbound tannins, and hydrolyzable tannins. To hydrolyze tannic acid into gallic acid, an acid, an alkali, or a microbial tannase can be used. Gallic acid and gallotannins are readily separated by oxidation. The quickest way to accomplish it is to use strong sulfuric acid to extract it out of an aqueous solution.

Gallic acid may be produced more slowly by allowing water to gradually oxidize in the presence of ambient oxygen. It is mostly used in the pharmaceutical industry to create antibacterial drugs like trimethoprim. Gallic acid is used as a substrate in the chemical synthesis of food preservatives such pyrogallol and gallates in the food industry (Mukherjee, Gargi, and Rintu Banerjee, 2003). Gallates are often recognized in a range of plants as the ester derivatives of gallic acid, and researchers are also looking into their biological characteristics.

The anticancer effects of Gallic acid ester derivatives have been discovered. These studies found that gallates induced apoptosis in a number of cancer cell types (Locatelli, Claudriana, Fabíola Branco Filippin-Monteiro, et al. 2013). According to structural activity relationship (SAR) study, gallic acid derivatives can act as an antioxidant since they have hydroxyl groups (Subramanian, A. P., A. A. John, 2015). The para-substituted-OH group possess a potent radical scavenger. Hydroxyl groups and intramolecular hydrogen bonds both have an impact on antioxidant activity. For instance, the readily ionizable carboxylic group in phenolic acids is responsible for their efficient hydrogen donation ability (Sroka, Z., and W. Cisowski. 2003). Gallic acid has an easily ionizable carboxyl group, which would lead to an effective hydrogen donating tendency of phenolic acids and is the main reason why it is more effective as an antioxidant than pyrogallo-5 (Badhani, Bharti, Neha Sharma, et al. 2015).

Due to its bioactivity and widespread availability in nature, gallic acid was consequently a crucial ingredient in the development of novel potent pharmacophores. Recent advances in the manufacture of gallic acid derivatives have led to the production of a large number of physiologically and pharmacologically active compounds. Apoptosis may be precisely triggered in cancer cells by gallic acid and its derivatives, without harming healthy cells, according to studies. Several physicochemical features, including the lipid solubility in the alkyl esters, 3,4,5-triacylated benzoic acid, and its esters, are due to chemical alterations done to the gallic acid molecule (Hejchman, Elżbieta, Przemysław Taciak, et al. 2015).

It has been shown that antioxidant compounds that are effective in both aqueous and lipid environments have improved in efficacy. Examples of this include the fact that a food additive called dodecyl gallate 7c inhibits the enzyme xanthine oxidase more potently than gallic acid. Oral *Streptococcus mutant* biofilms and bacterial development respond better to methyl gallate 7a than gallic acid as a format ion inhibitor. Nonyl gallate is more efficient than gallic acid at inhibiting the growth of *Salmonella choleraesuis* and preventing the development of skin malignancies induced by dimethyl benzanthracene. These biological processes have been linked to the amphipathic feature of these ester derivatives (Kubo, Isao, Ken-ichi Fujita, 2003) because it has antioxidant activity *in vitro* that is almost equal to gallic acid.

To discover its structure-activity correlations, gallic acid has experienced a variety of structural modifications. However, no attempt has been made to yet to link the biological effects of drugs that include analogues of gallic acid to their synthesis. This work aims to update the chemistry of gallic acid-containing pharmacophores and to investigate their medicinal potential. Information on gallic acid derivatives is categorized based on their chemical constitution (Subramanian, A. P., A. A. John, et al. 2015).

Different Pathways Involved in Mechanism of Action of Gallic Acid

The Tendency of Gallic Acid to Prevent DNA Methylation

Human cancers linked to tobacco use contain epigenome abnormalities. Through a suggested mechanism involving the gallate moiety of EGCG, epigallocatechin-3-gallate (EGCG) has been shown to alter gene expression by targeting DNA methyltransferases (DNMTs) (Paluszczak, Jarosław, Violetta Krajka-Kuźniak, et al. 2011). Gallic acid (GA) significantly lowers nuclear and cytoplasmic levels of DNMT1 and DNMT3B and modifies the methylome of lung cancer and pre-malignant oral cell lines. Compared to EGCG, gallic acid demonstrates greater cytotoxicity against the lung cancer cell line H1299. Because CCNE2 and CCNB1 are demethylated in H1299 cells, gallic acid may restart the GADD45 signaling pathway, which causes growth inhibition and DNA damage.

It was discovered that a fungus called *Aspergillus sojae* may effectively boost the gallic acid content in oolong tea during the fermentation process, improving the epigenetic anti-cancer effects of oolong tea. In the post-fermentation oolong tea extract (PFOTE), the fungus significantly elevated gallic acid, which boosted the demethylation effects and significantly decreased the nuclear abundances of DNMT1, DNMT3A, and DNMT3B in lung cancer cell lines. In H1299 cells, post-fermentation oolong tea extract (PFOTE) shown enhanced sensitivity to cisplatin and higher anti-proliferation effects than oolong tea extract (OTE). Gallic acid's substantial inhibitory effects on DNMT activities provide a solid scientific basis for the use of specialized fermented oolong tea high in gallic acid as an efficient dietary intervention method for malignancies linked to tobacco smoking (Weng, Yui-Ping, Pin-Feng Hung, et al. 2018).

Gallic Acid Accelerates EGFR Turnover, Promoting Apoptosis in EGFR-Mutant Non-Small Cell Lung Tumors

Non-small cell lung cancer (NSCLC) cells with the EGFR mutation undergo apoptosis in response to gallic acid, but not EGFR-WT NSCLC cells. Only in EGFR-mutant NSCLC cells on treatment with gallic acid cause a substantial decrease in proliferation and the activation of apoptosis (Wang, Dong, and Burenbatu Bao. 2020). It's significant to note that therapy with gallic acid resulted in a significant drop in EGFR levels, a key factor in NSCLC survival. Gallic acid therapy enhanced EGFR turnover but had no effect on transcription. In fact, therapy with a proteasome inhibitor significantly corrected the EGFR downregulation induced by gallic acid. Gallic acid also promoted EGFR turnover, which in turn encourages apoptosis in EGFR-TKI (tyrosine kinase inhibitor)-resistant cell lines, which depend on EGFR signaling for survival (Nam, Boas, Jin Kyung Rho, et al. 2016).

Through the JAK/STAT3 Signaling Pathway, Gallic Acid Possesses Anticancer Action and Amplifies the Anticancer Effects of Cisplatin in Non-Small Cell Lung Cancer A549 Cells

Gallic acid, which elevated the B-cell lymphoma 2 (Bcl2) associated X protein (Bax) and significantly suppressed Bcl2, enabled NSCLC A549 cells to undergo apoptosis and decrease their ability to proliferate in a dose and time-dependent manner. Notably, the findings also showed that gallic acid increased the anticancer effects of cisplatin in the prevention of growth of cancer cells and the activation of cell death after enhanced Bax expression and lowered Bcl2 expression. Additionally, the findings of this study showed that gallic acid had independent anticancer effects on NSCLC A549 cells and that it enhanced cisplatin's anticancer effects via modifying the JAK/STAT3 signaling pathway and downstream apoptotic molecules. The anticancer properties of gallic acid as well as its supporting effects on cisplatin action in individual NSCLC may be the subject of more fundamental research and preclinical trials (Zhang, Tingxiu, Lijie Ma, et al. 2019).

The Inhibitory Pathways of Tumor Cell PD-L1 Expression in Non-Small-Cell Lung Cancer (NSCLC) Cells by Natural Bioactive Gallic Acid

By binding growth factors like EGF, the epidermal growth factor receptor (EGFR) is phosphorylated, which activates downstream prooncogenic signaling pathways such KRAS-ERK, JAK-STAT, and PI3K-AKT (Huang, Yao, and Yongchang Chang. 2011). By stimulating uncontrollable cell cycle, multiplication, invasion, and expression of programmed death-ligand 1 (PD-L1), these pathways aid in the development of NSCLC tumors (Braun, Mitchell W., 2016). Although the treatment of NSCLC has advanced significantly thanks to new cytotoxic medicines, side effects remain a major factor in death.

The phenolic natural chemical gallic acid (3,4,5-trihydroxybenzoic acid; gallic acid) was isolated from plant derivatives and has been shown to have anticancer properties. Studies have looked at the tumor-suppressing properties of gallic acid, which reduced PD-L1 expression in NSCLC through binding to EGFR. This interaction prevented EGFR from being phosphorylated, which in turn prevented PI3K and AKT from being phosphorylated, leading to the activation of p53. PD-L1 was downregulated as a result of miR-34a's p53-dependent overexpression.

Research demonstrated the synergistic effects of gallic acid and anti-PD-1 monoclonal antibody in a co - culture system of NSCLC cells and peripheral blood mononuclear cells (Kang, Dong Young, Nipin Sp, et al. 2020). In advanced non-small cell lung cancer, gallic acid suppression of Src-Stat3 signaling overcomes acquired resistance to EGF receptor tyrosine kinase inhibitors.

Patients with lung cancer who carry a subset of functional EGFR mutations have had therapeutic improvement from tyrosine kinase inhibitors (TKIs) that target the EGFR. Gallic acid (GA), a naturally occurring polyphenolic molecule, exhibits anti-tumorigenic properties in TKI-resistant non-small cell lung cancer (NSCLC) (Mitsudomi, Tetsuya, and Yasushi Yatabe. 2007). The TKI-resistant cancer was more responsive to the tumor-suppressing effects of gallic acid than the TKI-sensitive cancer. Gallic acid administration caused TKI-resistant lung cancer to undergo apoptosis and cell cycle arrest, but not TKI-sensitive lung cancer, because it blocked Src-Stat3-mediated signaling and reduced the expression of genes that promote tumor growth.

Gallic acid-treated tumors that were separated from the xenograft model showed tumor-selective growth inhibition that was resistant to TKIs and minimization of Src-Stat3-dependent signaling. This discovery revealed the significance of the Src-Stat3 signaling cascade in the tumor-suppressive action of gallic acid and, more critically, offers a brand-new therapeutic perspective on the use of gallic acid for advanced TKI-resistant lung cancer (Phan, Ai NH, Tuyen NM Hua, et al. 2016).

Through the Amplification of Glutathione Depletion, MAPK Inhibitors Increase the Mortality of A549 Lung Cancer Cells Caused by Gallic Acid

Instead of having pro-apoptotic consequences, ERK activation has a pro-survival role. Similar to this, MEK inhibitor, which presumably inhibited ERK activity, greatly exacerbated cell death and somewhat improved the growth inhibition by gallic acid (Park, Woo Hyun. 2011). The percentage of A549 control cells was also dramatically decreased by MEK inhibitor alone. As a result, in gallic acid-treated A549 cells, MEK inhibitor's suppression of ERK signaling promotes apoptosis while acting as a growth inhibitor in A549 control cells.

An MEK inhibitor, however, lessens growth inhibition and cell death in Calu-6 cells that have been exposed to gallic acid, (Henson, Elizabeth S., et al. 2006) demonstrating that MEK inhibitor influences cell growth and death differently in gallic acid-treated A549 and Calu-6 lung cancer cells. Additionally, in human pulmonary fibroblast cells treated with gallic acid, MEK inhibitor had no effect on cell death or growth inhibition. As a result, while treating lung cancer, the targeted therapy linked to ERK signaling should be carefully evaluated.

The amount of CMF (GSH) in A549 cells was raised by gallic acid. The elevated GSH level was probably brought on by the rising ROS level caused by gallic acid. It's possible that certain cells that were unable to resist the oxidative stress brought on by treatment with gallic acid proceeded through a cell death pathway. In A549 cells treated with gallic acid, GSH levels were increased by all MAPK inhibitors.

These studies demonstrate that GSH levels are upregulated in gallic acid-treated A549 cells by MAPK inhibitors, which has an impact on ROS levels and the percentages of GSH-depleted cells. In A549 cells, gallic acid triggered growth inhibition, cell death, an increase in intracellular ROS, and the depletion of GSH. In gallic acid-treated A549 cells, all MAPK inhibitors increased growth inhibition and mortality, which were linked to GSH depletion rather than ROS level (Park, Woo Hyun, and Suhn Hee Kim. 2013).

Apoptosis Induced in H446 Cell Line Through the ROS-Dependent Mitochondrial Apoptotic Pathway

One of the biggest causes of cancer-related death globally is lung cancer. Additionally, SCLC is the most fatal subtype with early extensive distribution and rapid development as its distinguishing features. Although cisplatin and etoposide combination chemotherapy has been widely used to raise the recovery rate of SCLC, nearly all patients experience MDR recurring cancer and finally pass out from this condition (Wang, Ruixuan, Lijie Ma, et al. 2016).

Through an elevation in intracellular ROS, a decrease in MMP, an elevation in BAX, Apaf-1, DIABLO, and p53 expression, and a decrease in XIAP expression, gallic acid suppressed the development of SCLC H446 cells and increased the apoptosis inducing activities of cisplatin. These studies also suggest that gallic acid may be a useful adjunctive medication in the therapeutic management of SCLC, assisting in the reduction of cisplatin-induced MDR and toxicity. Programmed cell death, or apoptosis, is a strictly controlled process that is characterized by a number of morphological alterations.

The induction of apoptosis is known to be crucial in preventing the growth and spread of cancer, according to several studies. In H446 cells, gallic acid triggered apoptosis, changed the shape of the cells, and restricted their proliferation. These data demonstrate that the anticancer properties of gallic acid in H446 cells may be mediated through cell death (Prakash O. M., Amit Kumar, et al. 2013). ROS is a crucial mediator of intracellular signaling and is one of the by-products of typical cellular oxidative activities.

An increasing number of studies show that a greater concentration of ROS causes MMP breakdown, oxidation of cellular macromolecules, and ultimately cell death. The aberrant activation and expression of the Bcl-2 family, which includes anti-apoptotic molecules like Bcl-2 and Bcl-xL and pro-apoptotic proteins like Bak and Bax, is primarily responsible for mitochondrial-mediated cell death (Pawlak, Aleksandra, Witold Gładkowski, et al. 2018).

Conclusion

One of the most prevalent kinds of cancer, lung cancer has a high death rate and is known for being highly susceptible to chemo- and/or radiation therapy resistance. Over 80% of lung cancer cases globally are NSCLC, making it the most prevalent kind of cancer. Therefore, it is crucial to find innovative medications with little to no negative impacts.

Due to their possible anticancer properties and minimal toxicity, plant-derived chemicals have recently gained more and more public interest. Interestingly, gallic acid is a common phenolic molecule occurring in gallnuts, oak bark, sumac, grapes, and tea leaves, has been proven to be a powerful antioxidant medication and has also been demonstrated to contain excellent anti-tumor function in many cancer cells, including those of lung malignancies.

The chemical mechanism behind gallic acid's functions, however, is still completely understood and the research of gallic acid is insufficient. Therefore, the purpose of this study is to learn more about the role gallic acid plays in lung cancer. The gathered information revealed that gallic acid may potentially act on lung cancer's molecular targets.

References

Abdel-Razeq, Hikmat, Fadwa Attiga, and Asem Mansour. "Cancer care in Jordan." *Hematology/Oncology and Stem Cell Therapy* 8, no. 2 (2015): 64-70.

Ashrafizadeh, Milad, Ali Zarrabi, Sepideh Mirzaei, Farid Hashemi, Saeed Samarghandian, Amirhossein Zabolian, Kiavash Hushmandi et al. "Gallic acid for cancer therapy: Molecular mechanisms and boosting efficacy by nanoscopical delivery." *Food and Chemical Toxicology* 157 (2021): 112576.

Badhani, Bharti, Neha Sharma, and Rita Kakkar. "Gallic acid: a versatile antioxidant with promising therapeutic and industrial applications." *Rsc Advances* 5, no. 35 (2015): 27540-27557.

Barta, Julie A., Charles A. Powell, and Juan P. Wisnivesky. "Global epidemiology of lung cancer." *Annals of Global Health* 85, no. 1 (2019).

Braun, Mitchell W., and Tomoo Iwakuma. "Regulation of cytotoxic T-cell responses by p53 in cancer." *Translational Cancer Research* 5, no. 6 (2016): 692.

Brglez Mojzer, Eva, Maša Knez Hrnčič, Mojca Škerget, Željko Knez, and Urban Bren. "Polyphenols: Extraction methods, antioxidative action, bioavailability and anticarcinogenic effects." *Molecules* 21, no. 7 (2016): 901.

Choubey, Sneha, Lesley Rachel Varughese, Vinod Kumar, and Vikas Beniwal. "Medicinal importance of gallic acid and its ester derivatives: a patent review." *Pharmaceutical Patent Analyst* 4, no. 4 (2015): 305-315.

Choubey, Sneha, Soniya Goyal, Lesley R. Varughese, Vinod Kumar, Anil K. Sharma, and Vikas Beniwal. "Probing gallic acid for its broadspectrum applications." *Mini Reviews in Medicinal Chemistry* 18, no. 15 (2018): 1283-1293.

Denisenko, Tatiana V., Inna N. Budkevich, and Boris Zhivotovsky. "Cell death-based treatment of lung adenocarcinoma." *Cell Death & Disease* 9, no. 2 (2018): 1-14.

Fitzmaurice, Christina, Christine Allen, Ryan M. Barber, Lars Barregard, Zulfiqar A. Bhutta, Hermann Brenner, Daniel J. Dicker et al. "Global, regional, and national cancer incidence, mortality, years of life lost, years lived with disability, and disability-adjusted life-years for 32 cancer groups, 1990 to 2015: a systematic analysis for the global burden of disease study." *JAMA Oncology* 3, no. 4 (2017): 524-548.

Golemis, Erica A., Paul Scheet, Tim N. Beck, Eward M. Scolnick, David J. Hunter, Ernest Hawk, and Nancy Hopkins. "Molecular mechanisms of the preventable causes of cancer in the United States." *Genes & Development* 32, no. 13-14 (2018): 868-902.

Gredner, Thomas, Ute Mons, Tobias Niedermaier, Hermann Brenner, and Isabelle Soerjomataram. "Impact of tobacco control policies implementation on future lung cancer incidence in Europe: An international, population-based modeling study." *The Lancet Regional Health-Europe* 4 (2021): 100074.

Gurib-Fakim, Ameenah. "Medicinal plants: traditions of yesterday and drugs of tomorrow." *Molecular Aspects of Medicine* 27, no. 1 (2006): 1-93.

Hashim, Dana, and Paolo Boffetta. "Occupational and environmental exposures and cancers in developing countries." *Annals of Global Health* 80, no. 5 (2014): 393-411.

Hejchman, Elżbieta, Przemysław Taciak, Sebastian Kowalski, Dorota Maciejewska, Agnieszka Czajkowska, Julia Borowska, Dariusz Śladowski, and Izabela Młynarczuk-Biały. "Synthesis and anticancer activity of 7-hydroxycoumarinyl gallates." *Pharmacological Reports* 67, no. 2 (2015): 236-244.

Henson, Elizabeth S., and Spencer B. Gibson. "Surviving cell death through epidermal growth factor (EGF) signal transduction pathways: implications for cancer therapy." *Cellular Signalling* 18, no. 12 (2006): 2089-2097.

Huang, Yao, and Yongchang Chang. "Epidermal growth factor receptor (EGFR) phosphorylation, signaling and trafficking in prostate cancer." *Prostate Cancer-from Bench to Bedside* 143 (2011): 72.

Jain, Anuj, Gaurav Phull, and Sanjay K. Tiwari. "Management of severe Covid-19 cases through Ayurveda: A Case series." *International Journal of Ayurvedic Medicine* 13, no. 1: 191-198.

Kang, Dong Young, Nipin Sp, Eun Seong Jo, Alexis Rugamba, Dae Young Hong, Hong Ghi Lee, Ji-Seung Yoo, Qing Liu, Kyoung-Jin Jang, and Young Mok Yang. "The inhibitory mechanisms of tumor PD-L1 expression by natural bioactive gallic acid in non-small-cell lung cancer (NSCLC) cells." *Cancers* 12, no. 3 (2020): 727.

Kubo, Isao, Ken-ichi Fujita, Ken-ichi Nihei, and Noriyoshi Masuoka. "Non-antibiotic antibacterial activity of dodecyl gallate." *Bioorganic & Medicinal Chemistry* 11, no. 4 (2003): 573-580.

Locatelli, Claudriana, Fabíola Branco Filippin-Monteiro, and Tânia Beatriz Creczynski-Pasa. "Alkyl esters of gallic acid as anticancer agents: a review." *European Journal of Medicinal Chemistry* 60 (2013): 233-239.

Miller, Kimberly D., Leticia Nogueira, Theresa Devasia, Angela B. Mariotto, K. Robin Yabroff, Ahmedin Jemal, Joan Kramer, and Rebecca L. Siegel. "Cancer treatment and survivorship statistics, 2022." *CA: A Cancer Journal for Clinicians* 72, no. 5 (2022): 409-436.

Mitsudomi, Tetsuya, and Yasushi Yatabe. "Mutations of the epidermal growth factor receptor gene and related genes as determinants of epidermal growth factor receptor tyrosine kinase inhibitors sensitivity in lung cancer." *Cancer Science* 98, no. 12 (2007): 1817-1824.

MUKHERJEE, Gargi, and Rintu BANERJEE. "Production of gallic acid. Biotechnological routes (Part 2)." *Chimica Oggi* 21, no. 3-4 (2003): 70-73.

Nam, Boas, Jin Kyung Rho, Dong-Myung Shin, and Jaekyoung Son. "Gallic acid induces apoptosis in EGFR-mutant non-small cell lung cancers by accelerating EGFR turnover." *Bioorganic & Medicinal Chemistry Letters* 26, no. 19 (2016): 4571-4575.

Paluszczak, Jarosław, Violetta Krajka-Kuźniak, Zuzanna Małecka, Małgorzata Jarmuż, Magdalena Kostrzewska-Poczekaj, Reidar Grenman, and Wanda Baer-Dubowska. "Frequent gene hypermethylation in laryngeal cancer cell lines and the resistance to demethylation induction by plant polyphenols." *Toxicology in Vitro* 25, no. 1 (2011): 213-221.

Park, Woo Hyun, and Suhn Hee Kim. "MAPK inhibitors augment gallic acid-induced A549 lung cancer cell death through the enhancement of glutathione depletion." *Oncology Reports* 30, no. 1 (2013): 513-519.

ParK, Woo Hyun. "MAPK inhibitors differentially affect gallic acid-induced human pulmonary fibroblast cell growth inhibition." *Molecular Medicine Reports* 4, no. 1 (2011): 193-197.

Patil, P. and S. Killedar (2021). "Formulation and characterization of gallic acid and quercetin chitosan nanoparticles for sustained release in treating colorectal cancer." *Journal of Drug Delivery Science and Technology* 63: 102523.

Patil, P. and S. Killedar (2022). "Green Approach Towards Synthesis and Characterization of GMO/Chitosan Nanoparticles for In Vitro Release of Quercetin: Isolated from Peels of Pomegranate Fruit." *Journal of Pharmaceutical Innovation* 17(3): 764-777.

Pawlak, Aleksandra, Witold Gładkowski, Justyna Kutkowska, Marcelina Mazur, Bożena Obmińska-Mrukowicz, and Andrzej Rapak. "Enantiomeric trans β-aryl-δ-iodo-γ-lactones derived from 2, 5-dimethylbenzaldehyde induce apoptosis in canine lymphoma cell lines by downregulation of anti-apoptotic Bcl-2 family members Bcl-xL and Bcl-2." *Bioorganic & Medicinal Chemistry Letters* 28, no. 7 (2018): 1171-1177.

Phan, Ai NH, Tuyen NM Hua, Min-Kyu Kim, Vu TA Vo, Jong-Whan Choi, Hyun-Won Kim, Jin Kyung Rho, Ki Woo Kim, and Yangsik Jeong. "Gallic acid inhibition of Src-Stat3 signaling overcomes acquired resistance to EGF receptor tyrosine kinase inhibitors in advanced non-small cell lung cancer." *Oncotarget* 7, no. 34 (2016): 54702.

Prakash, O. M., Amit Kumar, and P. Kumar. "Anticancer potential of plants and natural products." *Am J Pharmacol Sci* 1, no. 6 (2013): 104-115.

Schultz, Jack C., Mark D. Hunter, and Heidi M. Appel. "Antimicrobial activity of polyphenols mediates plant-herbivore interactions." In *Plant Polyphenols*, pp. 621-637. Springer, Boston, MA, 1992.

Sroka, Z., and W. Cisowski. "Hydrogen peroxide scavenging, antioxidant and anti-radical activity of some phenolic acids." *Food and Chemical Toxicology* 41, no. 6 (2003): 753-758.

Štěpánek, Ladislav, Jarmila Ševčíková, Dagmar Horáková, Mihir Sanjay Patel, and Radka Durďáková. "Public Health Burden of Secondhand Smoking: Case Reports of Lung Cancer and a Literature Review." *International Journal of Environmental Research and Public Health* 19, no. 20 (2022): 13152.

Subramanian, A. P., A. A. John, M. V. Vellayappan, A. Balaji, S. K. Jaganathan, Eko Supriyanto, and Mustafa Yusof. "Gallic acid: prospects and molecular mechanisms of its anticancer activity." *Rsc Advances* 5, no. 45 (2015): 35608-35621.

Subramanian, A. P., A. A. John, M. V. Vellayappan, A. Balaji, S. K. Jaganathan, Eko Supriyanto, and Mustafa Yusof. "Gallic acid: prospects and molecular mechanisms of its anticancer activity." *Rsc Advances* 5, no. 45 (2015): 35608-35621.

Thandra, Krishna Chaitanya, Adam Barsouk, Kalyan Saginala, John Sukumar Aluru, and Alexander Barsouk. "Epidemiology of lung cancer." *Contemporary Oncology/ Współczesna Onkologia* 25, no. 1 (2021): 45-52.

Travis, William D., Elisabeth Brambilla, Andrew G. Nicholson, Yasushi Yatabe, John HM Austin, Mary Beth Beasley, Lucian R. Chirieac et al. "The 2015 World Health Organization classification of lung tumors: impact of genetic, clinical and radiologic advances since the 2004 classification." *Journal of Thoracic Oncology* 10, no. 9 (2015): 1243-1260.

Verma, Sharad, Amit Singh, and Abha Mishra. "Gallic acid: molecular rival of cancer." *Environmental Toxicology and Pharmacology* 35, no. 3 (2013): 473-485.

Wang, Dong, and Burenbatu Bao. "Gallic acid impedes non-small cell lung cancer progression via suppression of EGFR-dependent CARM1-PELP1 complex." *Drug Design, Development and Therapy* 14 (2020): 1583.

Wang, Ruixuan, Lijie Ma, Dan Weng, Jiahui Yao, Xueying Liu, and Faguang Jin. "Gallic acid induces apoptosis and enhances the anticancer effects of cisplatin in human small cell lung cancer H446 cell line via the ROS-dependent mitochondrial apoptotic pathway." *Oncology Reports* 35, no. 5 (2016): 3075-3083.

Weng, Yui-Ping, Pin-Feng Hung, Wen-Yen Ku, Chang-Yuan Chang, Bo-Han Wu, Ming-Han Wu, Jau-Ying Yao, Ji-Rui Yang, and Chia-Huei Lee. "The inhibitory activity of gallic acid against DNA methylation: Application of gallic acid on epigenetic therapy of human cancers." *Oncotarget* 9, no. 1 (2018): 361.

Xu, Xing, Mahdi Fallah, Yu Tian, Trasias Mukama, Kristina Sundquist, Jan Sundquist, Hermann Brenner, and Elham Kharazmi. "Risk of invasive prostate cancer and prostate cancer death in relatives of patients with prostatic borderline or in situ neoplasia: A nationwide cohort study." *Cancer* 126, no. 19 (2020): 4371-4378.

Zhang, Tingxiu, Lijie Ma, Pengfei Wu, Wei Li, Ting Li, Rui Gu, Xiaoping Dan, Zhiwei Li, Xianming Fan, and Zhenliang Xiao. "Gallic acid has anticancer activity and enhances the anticancer effects of cisplatin in non-small cell lung cancer A549 cells via the JAK/STAT3 signaling pathway." *Oncology Reports* 41, no. 3 (2019): 1779-1788.

Chapter 4

A New Perspective on the Efficacy of Gallic Acid Nanoformulation on Colorectal Cancer Treatment

Venketesh Kumbhar[*]
Suraj Tarihalkar[†]
Poournima Sankpal[‡]
and Sachinkumar Patil[§]

Ashokrao Mane College of Pharmacy,
Peth-Vadgaon, Kolhapur, Maharashtra, India

Abstract

Cancer is a broad collection of illnesses that can begin in practically any organ or tissue of the body. These illnesses are brought on when abnormal cells grow out of control, cross their normal boundaries to infect nearby body parts and/or spread to other organs. An estimated 10 million deaths were attributed to cancer in 2020, making it the second highest cause of death worldwide. Men are more likely to develop lung, prostate, colorectal, stomach, and liver cancer than women, who are more likely to develop breast, colorectal, lung, cervical, and thyroid cancer.

A polyphenolic substance called gallic acid (GA) has been shown to prevent a number of disorders. In addition to prescription medications, nutraceuticals and medicinal foods are being used to treat neurological

---

[*] Corresponding Author's Email: venketeshkumbhar58649@gmail.com.
[†] Corresponding Author's Email: tarihalkarsuraj262@gmail.com.
[‡] Corresponding Author's Email: poournima6@gmail.com.
[§] Corresponding Author's Email: sachinpatil.krd@gmail.com.

In: The Chemistry of Gallic Acid and Its Role in Health and Disease
Editor: Jeff C. Murdoch
ISBN: 979-8-88697-672-4
© 2023 Nova Science Publishers, Inc.

disorders, cancer, hepatitis C, inflammation, and cardiovascular diseases, even though the majority of these conditions have only been investigated *in vitro*. Colorectal cancer, which is the third most common disease worldwide and one of the main reasons why people die from cancer.

This chapter is based on research into various gallic acid nanoformulations, which illustrates their treatment resistance for colorectal cancer. The application of nanotechnology in medicine has had a significant impact on the development of theranostic agents, which may simultaneously diagnose and treat illnesses. Several nanocarriers have been created and successfully employed to carry pharmaceutical drugs, including graphene oxide, polymers-based delivery systems, layered double hydroxides, gold nanoparticles, multifunctional nanoparticles and iron oxide magnetite nanoparticles.

Keywords: gallic acid (GA), nanoformulation, colorectal cancer, thernostic agents

Introduction

Cancer is a significant public health concern on a global scale. Due to global demographic trends, it is predicted that by 2025, there will be more than 20 million new cases of cancer per year. According to GLOBOCAN data, 10.3 million cancer-related deaths and 19.3 million new cases are anticipated in 2022. The most common cancers diagnosed are colon, lung, and prostate, and female breast cancers. The most common and deadly kind of cancer worldwide is still lung cancer [1]. One to two million new cases of colorectal cancer (CRC) are diagnosed each year, making it one of the most prevalent cancers in the world. CRC is also the third most common cancer and the fourth most common cause of cancer-related death. By gender, CRC is the second most common cancer in women (9.2%) and the third in men (10%) [2, 3].

Traditional treatments like radiotherapy and chemotherapy are used when the disease has progressed, but these treatments are intrusive and have serious side effects, including as the development of drug resistance. For the treatment of CRC, there are now major efforts being made to find and develop a nanoformulation based on natural substance [4]. Conventional chemotherapy's main drawback is its negative effects on the body because it cannot target cancer cells specifically while sparing surrounding healthy cells or causing problems like cardiac, hepatic, pulmonary, renal, and gastrointestinal toxicities by rapidly dividing healthy cells like those of the gastrointestinal tract, bone marrow, and hair follicles [5-7].

Drug delivery systems have several advantages over conventional chemotherapy, including improved drug solubility, targeted distribution at the site of the disease, sustained release for longer bioavailability, and reduced dosage needs [8-13]. Medicine has been significantly impacted by the development of theranostic drugs that can both diagnose and treat diseases [14]. Several nanocarriers have been created and successfully employed to carry pharmaceutical drugs, including graphene oxide, polymers-based delivery systems, layered double hydroxides, gold nanoparticles, multi-functional nanoparticles, and iron oxide magnetite nanoparticles. White tea, witch hazel, and other plants and foods contain gallic acid (3, 4, 5-trihydroxybenzoic acid), a bioactive substance with anticancer, anti-inflammatory, and antioxidant activities.

Gallic acid is also well known for its protective effects on healthy cells. Because of their unique characteristics, which include ease of preparation, scalable production, sustained release features, high encapsulation capacity, biocompatibility with healthy cells and tissues, ease of surface modification, and stable magnetic nature, they are crucial for cancer therapy. The present review is based on new perceptive on efficacy of gallic acid nanoformulation on colorectal cancer treatment.

Gallic Acid

Processed beverages like red wines and green teas include gallic acid (3, 4, 5-trihydroxybenzoic acid), a naturally occurring polyphenolic component. It can be found in the bark, roots, and leaves of many different types of plants, including berries (pomegranates and gall nuts) [15]. A wide range of hard wood plant species, such as oak (*Quercus robur*), chestnut (*Castanea sativa*), and many more, also contain it. Trihydroxycinnamic acid or caffeic acid can be used to create it from the amino acid phenylalanine [16].

Gallic acid, known to affect several pharmacological and biochemical pathways have strong antioxidant [17], anti-inflammatory [18], antimutagenic [19] and anticancer properties [20]. In the presence of metal ions, gallic acid is also known to have pro-oxidant properties that vary on concentration. Gallic acid's pro-oxidant qualities have been identified as an apoptosis inducer in cancer cell types. Studies have also confirmed that gallic acid has a preventive effect against chemically induced carcinogenesis [21].

Gallic acid, commonly known as trihydroxybenzoin acids 3, 4, and 5, is a crystalline substance that is either barely yellow or colourless. Its chemical formula is C7H6O5, and its molecular weight is 170.11954 g/mol. With a breakdown temperature of 235 to 240°C and a melting point of 210°C, carbon dioxide and carbon monoxide are produced. It has a log P of 0.70 (20°C), a pKa of 4.40, and a density of 1.69 kg/L. It is nearly insoluble in benzene, chloroform, ether petroleum, and alcohol. It is soluble in water, alcohol, ether, and glycerol (PUBCHEM, 2015).

Even though its biogenesis has not been fully elucidated, it is known that it originated in the shikimic acid pathway, a crucial process in the creation of secondary metabolites with an aromatic structure found in plants and specific microbes. This pathway creates coumarins, alkaloids, lignans, and polyphenols, which are all significant classes of substances. It begins with certain amino acids, specifically L-phenylalanine. There are three different paths that could be taken to produce it. According to the first, phenylalanine is first converted to caffeic acid, then to acids 3, 4, and 5 trihydroxycinnamic, and ultimately to gallic acid. Another explanation contends that since this is not present in nature, it is produced directly from the acid 3, 4, and 5 trihydroxycinnamic. Thus, the formation of protocatechuic acid from caffeic acid results in the production of side chains. According to a third hypothesis, 3-dehydroshikimic acid is produced from shikimic acid by the action of the enzyme dehydrogenase. Gallic acid is produced as a result of a spontaneous aromatization, which leads to this [22].

Recent Developments, Impact and Challenges in Nanoformulations

For use in medication delivery systems, nanoformulations of pharmaceuticals have caught the attention of numerous researchers. These nanoformulations are unique to the intended delivery site and improve the characteristics of conventional medications. The pharmaceutical industry is using nanoformulations such as dendrimers, polymeric nanoparticles, liposomes, nanoemulsions, and micelles to improve drug formulation.

There are several different synthesis techniques that can be used to create nanoformulations that transport medications to biological systems. The size and shape of the particulate formulation, the biochemical characteristics of the medication, and the targeted site all influence the choice of synthesis methods [23].

The pharmaceutical business uses a variety of nanoformulations, including dendrimers, polymers, liposomes, nano-emulsions, and micelles, for drug delivery. In order to construct globular, tree-like structures, dendrimers are synthetic, hyperbranched polymeric macromolecules with welldefined core, backbone, and multivalent periphery [24, 25]. By employing covalent conjugations or encapsulation within the cores, these dendrimers have the ability to transport a variety of medicines [26, 27].

Additionally, they can be functionalized to fit the characteristics of the delivery site, giving them a better choice to create a medication with many functions [28]. Drug delivery vehicles for polymeric nanoparticles such nanospheres and nanocapsules are also employed [29]. Polymeric nanoparticles enhance the water solubility of drugs [30].

Synthesis Routes of Different Nanoformulations

There are several syntheses methods for various nanoformulations. It heavily depends on the kind of program and its features. Various synthesis methods that are used for the nanoformulation.

Dendrimers Synthesis

Divergent growth, convergent growth, and other ways make up the three categories under which dendrimer synthesis techniques are categorised. In the divergent technique, dendrimer development starts at a core location. While in the convergent technique, a dendrimer is produced by the reaction of multiple dendrons with a multifunctional core. The last approaches are the growth of hypercores and branching monomers, double exponential growth, lego chemistry, and click chemistry. Grouped in the third category primarily to improve the convergent and divergent growth methods [31].

Utilizing a divergent growth strategy, phosphorus dendrimers with seven generations and up to 384 functional groups were created. By using this technique, linear dendritic poly (ester)-block-poly (ether)-block-poly (ester) ABA copolymers, aliphatic ester dendrimers, polyphenylene dendrites, and dendrimers based on siloxane and carbosiloxane were also produced. These dendrimers are challenging to make using the convergent method because, due to their chemical makeup, their coupling to polymers via the Williamson ether reaction or metal catalysed Tran's esterification is not suitable for dendrimer formation.

The main disadvantage of the divergent growth method is the need to utilise an excessive amount of reagents and the challenge of synthesizing uniformly shaped dendrimers. Despite these drawbacks, this synthesis technique helps in the mass production of high-generation dendrimers.

Divergent Growth Method

First, the reaction between the core and two or more moles of reagent that contains at least two protective branching sites promotes dendrimer production using the divergent growth approach. Later, a chemical step removes the protective coating to create a first-generation dendrimer. Repeat this procedure until the desired dendritic size is obtained.

Convergent Growth Method

Convergent growth technique begins by synthesizing the dendrimer's surface and eventually builds the core by joining surface units with additional monomers. Due to the step-by-step assembly of its constituent parts, the convergent synthesis might be referred to as the "organic chemist's approach" to dendrimer synthesis.

It was Frechet and his colleagues who developed the first poly (aryl ether) dendrimer using this technique. The convergent growth approach is also used to synthesis poly (aryl alkyne) dendrimers, poly (phenylene) dendrimers, poly (alkyl ester) dendrimers, poly (aryl alkene) dendrimers, and poly (alkyl ether) dendrimers. This review is based on the new perceptive on the efficacy of gallic acid nanoformulation on colorectal cancer treatment. There are several advancement has been done in gallic acid nanoformulation [32].

Magnetite Nanocomposite Formulation of Gallic Acid (Fe_3O_4-PEG-GA) Against Colon Cancer Cells

Several cancer cell lines, including lung cancer cell line (A549), breast cancer cell line (MCF-7) and colon cancer cell line, were examined using the free drug GA, empty nanocarrier ($Fe3O4$-PEG), and developed magnetite nanocomposite formulation (Fe3O4-PEG-GA) (HT-29). In this work, conducted the experiment on breast cancer cell (MCF-7) because the percentage drug loading of GA is different from what was previously reported in a prior study that had tested the nanocomposite (P-Fe3O4-PEG-GA) against breast cancer cell (MCF-7) (P-Fe3O4-PEG-GA). Chemotherapeutic agent doxorubicin has undergone significant research, and its IC50 values for the same cancer cell lines have been reported to be 0.33 0.03 g/mL for HT-29, MCF-7, and A549, respectively. These Doxorubicin IC50 values served as a guide for the study's positive control [33].

Gallic Acid and Quercetin Chitosan Nanoparticles for Sustained Release in Treating Colorectal Cancer

Using the complete extract of either fruit, this study specifically aims to evaluate the therapeutic role of natural biomolecules such gallic acid (phenolic) isolated from amla fruit (Emblica Officinalis) and quercetin (flavonoid) from pomegranate fruit (Punica Granatum) peels because of its potent antioxidant and anticancer characteristics, numerous investigations have shown that gallic acid alters a wide variety of pharmacological and biochemical processes [34].

A useful nutraceutical for the treatment of cancer, quercetin is now a hot research issue, according to the bibliometric analysis of the Web of Science. Gallic acid and quercetin were combined into an o/w nanoemulsion utilising a GMO/chitosan framework, with a few minor alterations [35]. In a brief, extracted gallic acid (100 mg) and quercetin (100 mg) were dissolved in molten GMO (2 g), and 12.5 ml of 0.1% poloxamer 407 was added after that. The mixture was then sonicated at 18 W for three minutes in a probe sonicator. Using a probe sonicator set to 16 W for 4 minutes, 12.5 ml of 2.4% chitosan solution was once more added to this emulsion drop by drop. After conducting 12 cycles of the High Pressure Homogenizer (HPH) at 15,000 psi to create the nanoemulsion, this phase was lyophilized for 48 hours with 2% mannitol as a cryoprotectant. The combined effect of two variables, A: Chitosan (1.2-3.6 gm) and B: Poloxamer 407 (0.05-0.15 gm), each at 2 levels and the potential 9 combinations of CS nanoparticles, were examined using the central composite design. Gallic acid and quercetin CS nanoparticles were used in this study's oral nanotherapeutic method to treat colorectal cancer in a preclinical DMH rat model.

To assemble a stable formulation and targeted delivery, bioadhesive chitosan nanoparticles were created using quercetin and water-insoluble gallic acid as anticancer agents. Collectively, these results demonstrated that both biomolecules confirmed *in vitro* and *in vivo* can be delivered orally for colorectal cancer, based on the chitosan platform [36].

Epigallocatechin-3-gallate Nanoparticles

Compared with the non-loaded drug revealed a potential antioxidant movement [37]. As GA concentrations rise, human colon tumour HCT-15 cells undergo necrosis or apoptosis, which results in cell death and slows cell proliferation. It is found to have an influence on how HCT-15 cells form colonies. The colon cancer cells exposed to GA had noticeable morphological changes, such as membrane blabbing and cell shrinkage, which is consistent with earlier research on GA-treated HCT-15 cells. Additionally, the viability of Caco-2 cells in colorectal cancer (CRC) can be diminished by GA and 3-O-methylgallic acid [38]. This decline is mostly due by GA and 3-O-methylgallic acid's limited ability to obstruct the cell cycle at the G0/G1phaseGA and 3-O-methylgallic acid's ability to inhibit the translation of NF-kB, AP-1, OCT-1, and STAT-1 is limited, and this partially mediates their anti-tumor effects [38].

Polymer Nanoparticles Composed with Gallic Acid-Grafted Chitosan and Bioactive Peptides

The trypan blue assay was used to measure the density of human intestine Caco-2 cancer cells, and the GA-g-CS-CPP nanoparticles and EGCG loaded GA-g-CS-CPP nanoparticles have been shown to have anticancer potential against colon cancer. Trypan blue dye, which could not be stained, could be excluded from the live cells. The trypan blue assay revealed that grafting GA onto CS considerably (p 0.01) enhanced the polymer nanoparticles' ability to block Caco-2 cell proliferation in the cell line, and that this ability increased when the GA grafting ratio was raised.

The GA-g-CSs with GA substitutions of 26.5 1.0 and 126.0 1.1 mg/g had corresponding IC50 values of 154.4 and 115.5 g/m. Viability of the Caco-2 cancer cell after 72 hours of exposure to EGCG and EGCG-loaded GA-g-CS nanoparticles. Very intriguingly, EGCG's effects on inhibiting cell development were strengthened when combined with GAg-CS nanoparticles, particularly for the sample with a GA grafting ratio of 126.0 1.1 mg/g (p 0.01). In terms of EGCG concentration, the IC50 values for EGCG and the EGCG loaded GA-g-CS nanoparticles with GA replacements of 26.5 mg/g and 126.0 mg/g were 86.6, 77.3, and 50.2 g/mL, respectively. The interaction of EGCG with the GA-g-CS-CPP nanoparticles could therefore result in synergistic cancer cell growth suppression actions against human colon cancer cells, it would be concluded [39].

Gallic Acid-Loaded Sodium Alginate Microspheres

Drug delivery devices called microspheres capture the active ingredient in a matrix. A controlled release profile preserves the loss of material and activity while still targeting the desired area. In formulations with various ratios, sodium alginate (NaAlg) was used to create microspheres that contained GA. The range of arrest efficiency was 11.26 to 72.64%. Release experiments were done at a pH of 7.4.GA/NaAlg ratio 1/8 was determined to be the ideal situation. The microspheres were found to inhibit GA and exhibit regulated scanning electron microscopy (SEM), differential scanning calorimetry (DSC), X-ray diffraction (XRD), and Fourier transform infrared spectrometer (FTIR) tests were utilised to analyse the microspheres [40]. On Caco-2 cells, GA maintained its anticancer effects, and the DPPH (2, 2-diphenyl-1-picrylhydrazyl) technique showed antioxidant activity.

The anticancer activity of GA and the GA-containing microspheres was evaluated on colon cancer cells using the MTT ((3-(4, 5-Dimethylthiazol2-yl) 2, 5-diphenyltetrazolium bromide) test) [41]. The dehydrogenase action in the mitochondria of living cells reduces the yellow tetrazolium salts, resulting in the formation of purple formazan. At a wavelength of 570 nm, the color shift brought about by this transformation can be detected spectrophotometrically. The ability of the tetrazolium ring to be broken solely by mitochondrial activity in living cells distinguishes between live and dead cells. Anticancer activity was examined by MTT testing of the Caco-2 cell line derived from colon epidermal adenocarcinoma cells. According to the investigation, it was discovered that the anticancer activity varied by dose and duration. The antioxidant and anticancer properties of GA have been demonstrated to be preserved by microencapsulation [42].

Chitosan and Glyceryl Monooleate Nanostructures Containing Gallic Acid Isolated from Amla Fruit

Gallic acid nanoparticles were produced by combining poloxamer 407, chitosan, and glyceryl monooleate (GMO) with high pressure homogenization and a probe sonicator [43]. The produced nanoparticles were characterised using studies on particle size, zeta potential, DSC, XRD, SEM, entrapment efficiency, loading content, *in vitro* release, and stability. They showed zeta potential of +24.2 mV, average size of 180.80.21 nm, and 76.80% encapsulation of gallic acid. The study should make it simpler to extract, design, and construct nanoparticles for the protection and sustained release of gallic acid, especially in the colonic region, as the cumulative *in vitro* drug release up to 24 hours was obtained at a rate of 77.16%.

It has been discovered that the nanoparticle system consisting of polymer and stabiliser employed to generate the glyceryl monooleate/chitosan nanostructure has an impact on the physicochemical properties of gallic acid. Long-lasting *in vitro* gallic acid release from nanoparticles, positive charge on particle with good value, and model hydrophobic drug with nano particle size range were all noteworthy results from the tests. The discovery and development of novel nanoformulations based on natural products has thus received a lot of attention, and the proposed carrier system may be effective for delivering gallic acid to the colorectal region [44].

Gum Arabic-Stabilized Gallic Acid Nanoparticles

Gallic acid (GA), a naturally occurring phenolic compound, typically has trouble exerting its therapeutic benefits due to its rapid metabolism and excretion. In order to increase GA's bioavailability, gum arabic was used to encapsulate it into nanoparticles. Before they were subjected to a test, the synthesised nanoparticles (GANPs) were analysed for size, physicochemical properties, and antioxidant and antihypertensive effects using a number of well-known *in vitro* assays, such as the 1, 1-diphenyl-2-picrylhydrazyl (DPPH), nitric oxide scavenging (NO), carotene bleaching, and angiotensin-converting enzyme (ACE).

Four different human cancer cell lines, including MCF-7, MDA-MB231, hepatocellular carcinoma, colorectal cancer, HT-29, and breast epithelial cell line MCF-10A, were used to test the *in vitro* cytotoxicity, cell uptake, and cell migration of the GANPs. With a preference for HepG2 and MCF7 cancer cells, the GANPs demonstrates strong antioxidant activities and promising anti-cancer properties in a dose-dependent manner. The majority of cancer cells were able to absorb GANPs, and these cells tended to gather near the cell's nucleus. The HepG2 and MCF7 cancer cells also had a considerably larger proportion of apoptosis and responded more favourably to gallic acid nanoparticle treatment in the cell migration assay.

This work is the first to confirm the synergistic effects of gum Arabic, The antioxidant qualities, by increasing the selectivity towards cancer cells and improving the stability of the gallic acid encapsulation. The synthesised nanoparticles were also remarkably non-toxic in normal cells. These findings imply that GANPs may have therapeutic potential for the development of stronger drugs that are likely to be precise in their targeting of cancer cells [45].

Conclusion

A wide variety of illnesses, inclusion cancer, can develop in virtually every organ or bodily tissue Aberrant cells that multiply uncontrollably, cross normal cell boundaries, and either spread to other organs or infect nearby body parts are the root cause of several diseases. Cancer was projected to be the second most common cause of death worldwide in 2020, accounting for 10 million deaths. Men are more likely to develop lung, prostate, colorectal, stomach, and liver cancer compared to women, who are more likely to develop breast, colorectal, lung, cervical, and thyroid cancer.

Gallic acid (GA), a polyphenolic compound, has been demonstrated to protect against a range of diseases. Even though the bulk of these ailments have only been studied *in vitro*, nutraceuticals and medicinal foods are being utilised in addition to prescription drugs to treat neurological disorders, cancer, hepatitis C, inflammation, and cardiovascular diseases. These nanoformulations are tailored to the desired delivery site and improve the qualities of conventional medications. The pharmaceutical industry uses nanoformulations such dendrimers, polymeric nanoparticles, liposomes, nanoemulsions, and micelles to improve medicine formulation. Nanoformulations that deliver drugs to biological systems can be made using a variety of different synthesis processes.

This chapter is based on a new knowledge of advances, namely the effectiveness of gallic acid nanoformulation in the treatment of colorectal cancer. The development of gallic acid nanoformulation has undergone several changes. Gallic acid and quercetin chitosan nanoparticles for sustained release in the treatment of colorectal cancer, gallic acid and quercetin chitosan nanoparticles against colon cancer cells, gallic acid-loaded sodium alginate microspheres, chitosan and glyceryl monooleate nanostructures containing gallic acid derived from amla fruit, and epigallocatechin-3-gallate nanoparticles are some examples of the nanomaterials.

References

[1] Bray, Freddie, Jacques Ferlay, Isabelle Soerjomataram, Rebecca L. Siegel, Lindsey A. Torre, and Ahmedin Jemal. "Global cancer statistics 2018: GLOBOCAN estimates of incidence and mortality worldwide for 36 cancers in 185 countries." *CA: A Cancer Journal for Clinicians* 68, no. 6 (2018): 394-424.

[2] Forman, David, Jacques Ferlay, B. W. Stewart, and C. P. Wild. "The global and regional burden of cancer." *World Cancer Report* 2014 (2014): 16-53.

[3] Mármol, Inés, Cristina Sánchez-de-Diego, Alberto Pradilla Dieste, Elena Cerrada, and María Jesús Rodriguez Yoldi. "Colorectal carcinoma: a general overview and future perspectives in colorectal cancer." *International Journal of Molecular Sciences* 18, no. 1 (2017): 197.

[4] Singh, Parul, and Abhishek Singh. "Ocular adverse effects of anti-cancer chemotherapy." *J. Cancer Ther. Res* 1, no. 5 (2012).

[5] Agustoni, Francesco, Marco Platania, Milena Vitali, Nicoletta Zilembo, Eva Haspinger, Valentina Sinno, Rosaria Gallucci, Filippo De Braud, and Marina Chiara Garassino. "Emerging toxicities in the treatment of non-small cell lung cancer: ocular disorders." *Cancer Treatment Reviews* 40, no. 1 (2014): 197-203.

[6] Usman, Muhammad Sani, Mohd Zobir Hussein, Sharida Fakurazi, and Fathinul Fikri Ahmad Saad. "Gadolinium-based layered double hydroxide and graphene oxide nano-carriers for magnetic resonance imaging and drug delivery." *Chemistry Central Journal* 11, no. 1 (2017): 1-10.

[7] Santhosh, K. K., M. Das Modak, and P. Paik. "Graphene oxide for biomedical applications." *J Nanomed Res* 5, no. 6 (2017): 1-6.

[8] Barahuie, Farahnaz, Dena Dorniani, Bullo Saifullah, Sivapragasam Gothai, Mohd Zobir Hussein, Ashok Kumar Pandurangan, Palanisamy Arulselvan, and Mohd Esa Norhaizan. "Sustained release of anticancer agent phytic acid from its chitosan-coated magnetic nanoparticles for drug-delivery system." *International Journal of Nanomedicine* 12 (2017): 2361.

[9] Li, Yingbo, Yan Wang, Liu Tu, Di Chen, Zhi Luo, Dengyuan Liu, Zhuang Miao, Gang Feng, Li Qing, and Shali Wang. "Sub-acute toxicity study of graphene oxide in the Sprague-Dawley rat." *International Journal of Environmental Research and Public Health* 13, no. 11 (2016): 1149.

[10] Chung, Chul, Young-Kwan Kim, Dolly Shin, Soo-Ryoon Ryoo, Byung Hee Hong, and Dal-Hee Min. "Biomedical applications of graphene and graphene oxide." *Accounts of Chemical Research* 46, no. 10 (2013): 2211-2224.

[11] Saifullah, Bullo, Mohd Zobir B. Hussein, and Samer Hasan Hussein Al Ali. "Controlled-release approaches towards the chemotherapy of tuberculosis." *International Journal of Nanomedicine* 7 (2012): 5451-5463.

[12] Lammers, Twan, Silvio Aime, Wim E. Hennink, Gert Storm, and Fabian Kiessling. "Theranostic nanomedicine." *Accounts of Chemical Research* 44, no. 10 (2011): 1029-1038.

[13] Zhou, Yang, Xiangxiang Jing, and Yu Chen. "Material chemistry of graphene oxide-based nanocomposites for theranostic nanomedicine." *Journal of Materials Chemistry B 5,* no. 32 (2017): *6451-6470.*

[14] Lim, Eun-Kyung, Eunji Jang, Kwangyeol Lee, Seungjoo Haam, and Yong-Min Huh. "Delivery of cancer therapeutics using nanotechnology." *Pharmaceutics* 5, no. 2 (2013): 294-317.

[15] Mohapatra, PK Das, Keshab C. Mondal, and Bikas R. Pati. "Production of tannase through submerged fermentation of tannin-containing plant extracts by Bacillus licheniformis KBR6." *J. Microbiol* 300 (2006): 297-301.

[16] Prince, Ponnian Stanely Mainzen, Hansi Priscilla, and Periathambi Thangappan Devika. "Gallic acid prevents lysosomal damage in isoproterenol induced cardiotoxicity in Wistar rats." *European Journal of Pharmacology* 615, no. 1-3 (2009): 139-143.

[17] Golumbic, Calvin, and H. A. Mattill. "The antioxidant properties of gallic acid and allied compounds." *Oil & Soap* 19, no. 8 (1942): 144-145.

[18] Kroes, B. H. vd, A. J. J. Van den Berg, H. C. Quarles Van Ufford, H. Van Dijk, and R. P. Labadie. "Anti-inflammatory activity of gallic acid." *Planta Medica* 58, no. 06 (1992): 499-504.

[19] Gichner, T., F. Pospíšil, J. Velemínský, V. Volkeová, and J. Volke. "Two types of antimutagenic effects of gallic and tannic acids towards N-nitroso-compounds-induced mutagenicity in the amesSalmonella assay." *Folia Microbiologica* 32, no. 1 (1987): 55-62.

[20] Mirvish, Sidney S., Antonio Cardesa, Lawrence Wallcave, and Philippe Shubik. "Induction of mouse lung adenomas by amines or ureas plus nitrite and by N-nitroso compounds: effect of ascorbate, gallic acid, thiocyanate, and caffeine." *Journal of the National Cancer Institute* 55, no. 3 (1975): 633-636.

[21] Senapathy, J. Giftson, S. Jayanthi, P. Viswanathan, P. Umadevi, and N. Nalini. "Effect of gallic acid on xenobiotic metabolizing enzymes in 1, 2-dimethyl hydrazine induced colon carcinogenesis in Wistar rats–a chemopreventive approach." *Food and Chemical Toxicology* 49, no. 4 (2011): 887-892.

[22] Fernandes, Felipe Hugo Alencar, and Hérida Regina Nunes Salgado. "Gallic acid: review of the methods of determination and quantification." *Critical Reviews in Analytical Chemistry* 46, no. 3 (2016): 257-265.

[23] Jeevanandam, Jaison, Yen San Chan, and Michael K. Danquah. "Nano-formulations of drugs: recent developments, impact and challenges." *Biochimie* 128 (2016): 99-112.

[24] Tomalia, Donald A., Adel M. Naylor, and William A. Goddard Iii. "Starburst dendrimers: molecular-level control of size, shape, surface chemistry, topology, and flexibility from atoms to macroscopic matter." *Angewandte Chemie International Edition in English* 29, no. 2 (1990): 138-175.

[25] Svenson, Sonke. "Dendrimers as versatile platform in drug delivery applications." *European Journal of Pharmaceutics and Biopharmaceutics* 71, no. 3 (2009): 445-462.

[26] Lee, Cameron C., John A. MacKay, Jean MJ Fréchet, and Francis C. Szoka. "Designing dendrimers for biological applications." *Nature Biotechnology* 23, no. 12 (2005): 1517-1526.

[27] Liu, Mingjun, Kenji Kono, and Jean MJ Fréchet. "Water-soluble dendritic unimolecular micelles: Their potential as drug delivery agents." *Journal of Controlled Release* 65, no. 1-2 (2000): 121-131.

[28] Wolinsky, Jesse B., and Mark W. Grinstaff. "Therapeutic and diagnostic applications of dendrimers for cancer treatment." *Advanced Drug Delivery Reviews* 60, no. 9 (2008): 1037-1055.

[29] Rao, J. Prasad, and Kurt E. Geckeler. "Polymer nanoparticles: Preparation techniques and size-control parameters." *Progress in Polymer Science* 36, no. 7 (2011): 887-913.

[30] Wu, Yunpeng, Yaguang Luo, and Qin Wang. "Antioxidant and antimicrobial properties of essential oils encapsulated in zein nanoparticles prepared by liquid–liquid dispersion method." *LWT-Food Science and Technology* 48, no. 2 (2012): 283-290.

[31] Nanjwade, Basavaraj K., Hiren M. Bechra, Ganesh K. Derkar, F. V. Manvi, and Veerendra K. Nanjwade. "Dendrimers: emerging polymers for drug-delivery systems." *European Journal of Pharmaceutical Sciences* 38, no. 3 (2009): 185-196.

[32] Matthews, Owen A., Andrew N. Shipway, and J. Fraser Stoddart. "Dendrimers—branching out from curiosities into new technologies." *Progress in Polymer Science* 23, no. 1 (1998): 1-56.

[33] Rosman, Raihana, Bullo Saifullah, Sandra Maniam, Dena Dorniani, Mohd Zobir Hussein, and Sharida Fakurazi. "Improved anticancer effect of magnetite nanocomposite formulation of gallic acid (Fe3O4-PEG-GA) against lung, breast and colon cancer cells." *Nanomaterials* 8, no. 2 (2018): 83.

[34] Babbar, Neha, Harinder Singh Oberoi, and Simranjeet Kaur Sandhu. "Therapeutic and nutraceutical potential of bioactive compounds extracted from fruit residues." *Critical Reviews in Food Science and Nutrition* 55, no. 3 (2015): 319-337.

[35] Senapathy, J. Giftson, S. Jayanthi, P. Viswanathan, P. Umadevi, and N. Nalini. "Effect of gallic acid on xenobiotic metabolizing enzymes in 1, 2-dimethyl hydrazine induced colon carcinogenesis in Wistar rats–a chemopreventive approach." *Food and Chemical Toxicology* 49, no. 4 (2011): 887-892.

[36] Patil Poounima, and Suresh Killedar. "Formulation and characterization of gallic acid and quercetin chitosan nanoparticles for sustained release in treating colorectal cancer." *Journal of Drug Delivery Science and Technology* 63 (2021): 102523.

[37] Ho, Hsieh-Hsun, Chi-Sen Chang, Wei-Chi Ho, Sheng-You Liao, Cheng-Hsun Wu, and Chau-Jong Wang. "Anti-metastasis effects of gallic acid on gastric cancer cells

involves inhibition of NF-κB activity and downregulation of PI3K/AKT/small GTPase signals." *Food and Chemical Toxicology* 48, no. 8-9 (2010): 2508-2516.

[38] Celep, A. Gulcin Sagdicoglu, Aslihan Demirkaya, and Ebru Kondolot Solak. "Antioxidant and anticancer activities of gallic acid loaded sodium alginate microspheres on colon cancer." *Current Applied Physics* (2020).

[39] Boppana, Rashmi, Raghavendra V. Kulkarni, G. Krishna Mohan, Srinivas Mutalik, and Tejraj M. Aminabhavi. "*In vitro* and *in vivo* assessment of novel pH-sensitive interpenetrating polymer networks of a graft copolymer for gastro-protective delivery of ketoprofen." *RSC Advances* 6, no. 69 (2016): 64344-64356.

[40] Larosa, Claudio, Marco Salerno, Juliana Silva de Lima, Remo Merijs Meri, Milena Fernandes da Silva, Luiz Bezerra de Carvalho, and Attilio Converti. "Characterisation of bare and tannase-loaded calcium alginate beads by microscopic, thermogravimetric, FTIR and XRD analyses." *International Journal of Biological Macromolecules* 115 (2018): 900-906.

[41] Gowdhami, Balakrishnan, Yesaiyan Manojkumar, R. T. V. Vimala, Venkatesan Ramya, Balakrishnan Karthiyayini, Balamuthu Kadalmani, and Mohammad Abdulkader Akbarsha. "Cytotoxic cobalt (III) Schiff base complexes: *in vitro* anti-proliferative, oxidative stress and gene expression studies in human breast and lung cancer cells." *BioMetals* 35, no. 1 (2022): 67-85.

[42] Monteiro-Riviere, N. A., A. O. Inman, and L. W. Zhang. "Limitations and relative utility of screening assays to assess engineered nanoparticle toxicity in a human cell line." *Toxicology and Applied Pharmacology* 234, no. 2 (2009): 222-235.

[43] Rafiee, Zahra, Mohammad Nejatian, Marjan Daeihamed, and Seid Mahdi Jafari. "Application of different nanocarriers for encapsulation of curcumin." *Critical Reviews in Food Science and Nutrition* 59, no. 21 (2019): 3468-3497.

[44] Mazzarino, Letícia, Heloísa da Silva Pitz, Ana Paula Lorenzen Voytena, Adriana Carla Dias Trevisan, Rosa Maria Ribeiro-Do-Valle, and Marcelo Maraschin. "Jaboticaba (Plinia peruviana) extract nanoemulsions: development, stability, and *in vitro* antioxidant activity." *Drug Development and Industrial Pharmacy* 44, no. 4 (2018): 643-651.

[45] Shukla, Shipra, Baljinder Singh, Arti Singh, and Charan Singh. "Emerging and Advanced Drug Delivery Systems for Improved Biopharmaceutical Attributes of Gallic Acid: A Review." *Phytomedicine Plus* (2022): 100369.

Chapter 5

Strategic Approaches of Gallic Acid in Different Disease Conditions

Komal Patekar[*]
Sakshi Patil
Poournima Sankpal
and Sachinkumar Patil
Ashokrao Mane College of Pharmacy,
Peth-Vadgaon, Kolhapur Maharashtra, India

Abstract

Gallic acid (GA) is a natural phenolic compound. It is found in plants and foods that have been esterified with sugars, also known as Gallo tannins. It is present in fruits, such as black currant, berries, grapes, avla, clove. It is said to have a number of health enhancing properties. It appears in plants as hydrolysable tannins, free acids, esters, catechin derivatives, and free acids. In numerous conditions where oxidative stress has been linked, such as cardiovascular diseases, cancer, neurological disorders, and ageing, it has been shown to have potential therapeutic and preventive effects.

In order to portray the pharmacological status of this compound for future studies, this chapter aims to summarize the pharmacological and biological activities of (GA), including anticancer, antioxidant, and anti-inflammatory, antidiabetic, anti-myocardial, anti-obesity, antimicrobial properties among this numerous data *in vitro* and animal models. The outcomes of this review may highlight the potential of this compound as a novel therapy approach for the conditions listed, either on its own or in

---

[*] Corresponding Author's Email: komalpatekar087@gmail.com.

In: The Chemistry of Gallic Acid and Its Role in Health and Disease
Editor: Jeff C. Murdoch
ISBN: 979-8-88697-672-4
© 2023 Nova Science Publishers, Inc.

combination with other medications or their analogues to enhance their effects.

Results of the present study demonstrate that GA and effective supplement, as an adjuvant therapy may be a very promising compound reducing mortality rate of diseases.

Keywords: gallic acid, anticancer, antioxidant, antimicrobial, antidiabetic, anti-obesity, anti-inflammatory, anti-myocardial

Introduction

Gallic acid, also known as 3,4,5-trihydroxy benzoic acid, is one of the most abundant phenolic acids in plants. It has strong antioxidant and free radical scavenging properties and can protect biological cells, tissues, and organs from oxidative stress damage. It's a colourless or slightly yellow crystalline compound that's commonly used in the food and pharmaceutical industries. It is a polyphenolic compound found naturally in processed beverages such as red wines and green teas [1]. It is a naturally existing secondary metabolite that has been isolated from a variety of fruits, plants, and nuts. According to pharmacokinetic studies, gallic acid absorbs and eliminates quickly after oral administration, and structural optimization or dosage form adjustment of gallic acid is beneficial to increase its absorption and elimination [2]. Many polyphenols gallic acid is a low molecular weight tri phenolic compound with excellent anti-inflammatory and antioxidative activities. In addition, gallic acid also has several evident pharmacological effect including anti-tumor, antibacterial, anti-diabetes, anti-obesity, antimicrobial and anti-myocardial ischemia [3].

Cancer is the second leading cause of death worldwide, accounting for an estimated 9.6 million deaths in 2018. Lung cancer is the most common cancer and the leading cause of cancer death in men, followed by prostate and colorectal cancer (in terms of incidence) and liver and stomach cancer (for mortality). Breast cancer is the most common cancer diagnosed in women and the common cause of cancer death, followed by colon cancer and lung cancer (in terms of incidence), and vice versa (for mortality). Cervical cancer is the fourth most common cancer type in terms of both incidence and mortality, affecting women worldwide. Cancer treatment with a solitary therapy is rarely effective because cancer development involves many aspects of the cell [4]. Combined treatment is a popular method of chemotherapeutics. Because different pathways will be specifically aimed and lower dosages of each individual drug will be required, the toxicity will be significantly reduced. A number of studies have shown that GA and its derivatives have anticancer activity, both *in vivo* and *in vitro*. GA has anticancer activity and improves cisplatin's anticancer effects in non-small cell lung cancer A549 cells via the JAK/STAT3 signaling pathway.

Oxidative stress, caused by an excess of free radicals, is the leading cause of many neurodegenerative, including cancer, atherosclerosis, cardiovascular disease, ageing, and inflammatory diseases. Polyphenols are a type of natural sources anti-oxidant with numerous biological activities, including antitumor activity, antifungal, antimicrobial, antiviral, antiulcer, and anti-cholesterol. GA derivatives have also been discovered in a variety of herbal drugs with diverse biological and pharmacological effects such as radical scavenging, interfering with cellular signaling pathways, and inducing cancer cell apoptosis. Because of their free radical scavenging and antioxidant properties, GA and its derived products such as lauryl gallate, propyl gallate, and hexadecyl gallate can prevent the oxidation and rancidity of oils and fats. As a result, they have the potential to be useful as food additives.

The uncontrolled use of antimicrobials has broadened the range of organisms' exposure. Diabetes (DM) is a prolonged metabolic disorder characterised by insulin dysfunction, resulting in hyperglycemia. It resulted in long-term organ damage, dysfunction, and failure. Diabetes mellitus is expected to affect 366 million people by 2030, according to the World Health Organization [3].

Diabetes mellitus (DM) is caused solely by a disruption in homeostasis in the host metabolism, which results in reactive hypoglycemia or insulin resistance. Diabetic patients are always associated with a chronic metabolic syndrome that manifests at structural, neurobiological, and psychological levels of presenter organs, appearing with organ injury diseases such as renal dysfunction, retinopathy, neuropathic pain, congestive heart failure, and angiopthy. Type 1 diabetes mellitus (DM) is distinguished by a loss of pancreatic insulin secretion, which is caused primarily by autoimmune pancreatic ß-cell destruction. Type 2 diabetes mellitus (DM), the most common type of diabetes (affecting approximately 90% of cases), is caused by insulin insensitivity in target tissues, which is accompanied by partial ß-cell dysfunction [3, 4].

Anticancer Activity

Despite considerable research efforts, there are no effective treatments for cancer, which is one of the world's deadliest diseases. Because cells in a healthy organism are programmed for coordination and cooperation under normal physiological conditions, cell disruption can result in a variety of deadly diseases, including cancer. Cancer is defined by an abnormal increase in cell division, cell death resistance in the generated cells, and a proclivity for invasion and metastasis [5].

Malignant cells disrupt the normal functions of other cells through invasion or metastasis. Cancer takes precedence over overall quality of life, irrespective of the cause, as according to official reports of the health and wellness organizations [6]. The individual and societal effects of cancer are quite large, and new medications are still being researched to control this issue [7]. By altering the balance between antioxidants and pro-oxidants, gallic acid can have both cytotoxic and antitumor effects [8]. By boosting the activity of superoxide dismutase (SOD), catalase (CAT), glutathione reductase (GR), and glutathione peroxidase (GPx), as well as by lowering lipid peroxidation and ROS production, the substance may in some situations be able to prevent the ROS-induced carcinogenesis. Gallic acid can also cause apoptosis, autophagy, and cell cycle arrest by activating the caspase pathway and generating ROS. By reducing the expression and activity of matrix metalloproteinases, it can also prevent invasion and metastasis [9].

It is possible to reduce the stages of tumour marker enzymes such as aspartate transaminase (AST), alanine aminotransferase (ALT), lactate dehydrogenase (LDH), alkaline phosphatase (ALP), and gamma-glutamyl transaminase by preventing the growth of liver cells (GGT) [10, 11].

Antioxidant Activity

The purpose of grafting amount (2016). Furthermore, in *in vivo* cancer models, gallic acid derivative products such as isobutyl gallate-3,5-dimethyl ether and methyl gallate-3,5-dimethyl ether can shrink tumours and improve survival rates [12]. Gallic acid regulates the expression of cell cycle proteins such as cyclin A, cyclin D1, and cyclin E, and it slows cell proliferation by stimulating the p27KIP enzyme and trying to suppress CDK activity [13] Gallic acid reduced tumour size in hepatocellular carcinoma, and the antioxidant gallic acid (GA) serum on to GN in situ gelling copolymers made of biodegradable gelatin and thermosensitive poly (N-isopropylacrylamide) for intracameral delivery of pilocarpine in antiglaucoma therapy [14]. More information indicates that GA content is increasing [15].

This research revealed that GA inhibits RS production and maintains a greater GSH/GSSG ratio in order to exert anti-melanogenic activity and antioxidant properties2. Gallic acid, ascorbic acid, and xanthone are three significant natural antioxidants. Naksuriya et al. (2015) studied the antioxidant capacity of curcumin in comparison with these three compounds on free radical scavenging action and the impact of their combination on this activity. The use of curcumin in combination with ascorbic acid or xanthone should be avoided, according to authors, who also indicate that curcumin-gallic acid is a potential antioxidant mixture that can be employed instead of the individual substance [14].

The phenolic content of medicinal plants is what gives them their antioxidant action readily available organic antioxidants (such as ascorbic acid and gallic acid) Gallic acid has historically been used as an antioxidant because it has significant oxygen-derived free radical scavenging action and because it has polyphenolic functionality [15]. Because of its antioxidant properties, it reduces the rancidity and spoilage of fats and oils, making it easier to use as a food ingredient in a variety of edible products like candy, chewing gum, and baked goods [16]. It can treat albuminuria and act as an antioxidant to prevent oxidative damage to human cells. Additionally, they stated that *T. bellerica* fruit are a rich source of gallic acid (2290 g/g dry weight of plant material) [17]. Polyphenol has been shown to have antioxidant action [18, 19].

Even in wines' low concentrations, polyphenol components were discovered to be effective free radical scavengers [20]. They acted in grapes or wines far more effectively than other commercially available inuria and diabetic treatments, as well as acting as a distant astringent in cases of internal bleeding [21]. Gallic acid has been shown to have a variety of vascular protective effects, including non-enzymatically oxidised gallic acid in physiological solutions that produces superoxide anions, low levels of H_2O_2 and activation of cyclooxygenase, endothelium-independent relaxation that is dependent on H_2O_2 levels and the activation of smooth muscle K+ channels, and an irreversible, slowly-developing endothelium-independent relaxation [21, 22].

Antimicrobial Activity

According to research on the structure-activity relationship of phenolic acids, a number of factors, including that of the basic chemical structure, the number, position, and substituents of the hydroxyl groups on the phenyl group, as well as the esterification of the carboxyl group, can affect antimicrobial activity [23, 24]. Hydroxycinnamic acids frequently have greater antimicrobial properties than hydroxybenzoic acids1 [25, 26]. When compared to their parent structures, hydroxybenzoic acids with lower levels of hydroxylation in phenol groups, highly methoxylated phenol groups, significantly oxidised phenol groups, or ester derivatives with long alkyl chains demonstrated stronger antibacterial activity [27]. In contrast, hydroxybenzoic acids with more unbound -OH groups on the phenol ring were found to be more effective against hepatitis C virus (HCV) and human immunodeficiency virus (HIV) [28, 29].

When exposed to gallic acid, *Pseudomonas aeruginosa*, *Staphylococcus aureus*, *Streptococcus* species, *Chromobacterium violaceum*, and *Listeria monocytogenes* all exhibit reduced motility, adhesion, and biofilm formation [30]. Furthermore, the substance has the ability to change the charge, hydrophobicity, and penetrability of the membrane surface in Gram-positive and Gram-negative bacteria5 [31, 32]. Gallic acid has been shown to increase antibiotic accumulation by decreasing membrane permeability in *Campylobacter jejuni* [33]. Furthermore, by chelating divalent cations, such as those produced by the hepatitis C virus (HCV) and the HIV virus, it can destroy the outer membrane of Gram-negative bacteria [34].

Gallic acid has been shown to inhibit bacterial dihydrofolate reductase and to stimulate topoisomerase IV-mediated DNA breakage in certain bacteria, in addition to its effects on the bacterial cell membrane [35]. Alkyl gallates can also enter bacterial cells, disrupting both cellular respiration and the electron transport chain [36]. Some gallic acid ester derivatives, such as octyl gallate, bind to the polar surface of the cell membrane utilising the hydrophilic catechol component as a hook and then enter the lipid bilayer using the hydrophobic alkyl part. They then function as a nonionic surfactant and obstruct the selective permeability of fungus cell membranes [37].

HCV adherence and absorption, HCV replication, HCV serine protease, HIV-1 integrase, HIV-1 transcriptase, and HIV-1 protease dimerization [38], as well as HSV-1 and HSV-2 attachment and penetration, can all be prevented by gallic acid. The particles of Haemophilus influenza A and B are also disrupted [36] Gallic acid can bind to glutamate-gated chloride channels in the nervous system of *Caenorhabditis elegans*, causing cell membranes to become hyperpolarized and muscles to contract in response to parasites. Finally, the worm is paralysed and dies [39] Gallic acid, alkyl gallates, and gallic acid formulations based on chitosan can help to boost the antibacterial action of other antibiotics such as erythromycin, gentamicin, norfloxacin, ciprofloxacin, ampicillin, penicillin, and oxacilli through a process known as synergism [37, 39].

Anti-Diabetic Activity

Gallic acid as a diabetes preventative one of the most significant public health issues, diabetes mellitus (DM), is a chronic metabolic illness that affects 346 million people worldwide [40]. It is primarily characterised by hyperglycemia, which is brought on by errors in insulin activity or secretion. Polydipsia, polyphagia, polyuria, weight loss, weariness, and eyesight loss have all been listed as symptoms of DM. These signs or even nonexistent, may be negligible [41].

Chronic diabetes mellitus is a major public health issue (DM). De Oliveira et al. review the effect of GA on biochemical and histological parameters as well as indicators of oxidative stress in the liver and kidney of streptozotocin (STZ)-induced diabetic rats [42]. Since GA lowers TG, TC, and LDL levels, it promotes a shift in the lipid profile of the control animals. In addition to lowering the amounts of free radicals and lipid peroxidation, it also increases the protection provided by enzymatic and non-enzymatic cancer prevention agents in these tissues. Their research shows that GA may be helpful in treating the hepatic and renal complications associated with DM and increases the possibility of another use as a comparable treatment for hypoglycemic medicines [43].

According to a study, the natural flavonoids GA and ellagic acid can be used as antidiabetic medications. Nair et al. examined the interaction between glycogen phosphorylase (GP), a crucial contender for maintaining glucose homeostasis, and GA and its dimer, ellagic acid [44]. According to reports, this dimer has a stronger inhibitory effect than GA. Both of these bioactive substances work as non-competitive inhibitors of the allosteric activator, AMP, and competitive inhibitors of glucose-1-phosphate. Because of this, ellagic acid has the potential to be employed as an antihyperglycemic agent [45].

Anti-Obesity Activity

A person's overall health and well-being are dependent on proper nutrition, physical activity, and maintaining a healthy body weight. Because our current obesogenic environment encourages increased food intake, non-healthy foods, and physical inactivity, weight gain and disease are natural and expected outcomes [45]. Natural products (nutraceuticals) are currently being studied on a large scale as potential treatments for obesity and weight management [46]. Gallic acid (GA), a natural phenolic acid with the chemical formula $C_6H_2(OH)_3COOH$, is a common secondary metabolite found in a wide variety of edible plants, including gall nut, tea leaf, oak bark, blueberry, grape seed, rose flower, sumac, witch hazel, *Syzygium cumini* fruit, and others [47]. The biological activities of GA and its derivatives may contribute to the nutritional value of these GA-rich edible plants. Tea, for example, has been shown in extensive research to have fat and antioxidant properties [48].

Tea's phenolic compounds have been shown to have a wide range of pharmacological activities, including antioxidant, anti-inflammation, and protective properties against severe metabolic disorders such as cardiovascular and neurodegenerative diseases, cancer, hyperlipidemia, obesity, and diabetes. White and green tea polyphenols rich in catechins and GA have been found to be abundant [46]. Green tea has 13.7 to 24.7 g of total polyphenols per 100 g, while white tea has 10.60 to 25.95 g/100 g. Green and white tea's main bioactive catechins are epigallocatechin gallate (EGCG), which is derived from epigallocatechin, and GA. Accumulating human evidence suggests that the antioxidant, anti-inflammatory, and mitochondrial function modulator EGCG is responsible for tea's anti-chronic disease properties [5]. Furthermore, GA and protocatechuic acids were discovered to be antihyperglycemic principles in *Hibiscus sabdariffa*, which is used as a refreshment food drink. Consumption of this drink has also been shown to improve hypertension, dyslipidemia, and liver disorders. The antioxidant and antiglycation properties of some edible plants, such as *Anacardium humile* St. Hill., have been linked to quercetin, catechin, and GA bioactivities, which may inhibit glycolytic enzyme activity in RAW264.7 macrophages. Ethnobotanical claims indicate that we have made progress in our understanding of bioactive components in traditional plants and their links to obesity. With the global prevalence of obesity increasing, nutraceuticals play an important role in its prevention and treatment [48].

Obesity is defined as an excess of body fat that frequently results in significant health impairment. It is the most common nutritional disorder and is considered a global epidemic. Obesity occurs when the size or number of fat cells in a person's body increases, potentially opening the door to a variety of chronic physical and psychological problems. It increases the risk of a variety of diseases, including heart disease, type 2 diabetes, obstructive sleep apnea, certain cancers, and osteoarthritis. It is most commonly caused by a combination of excessive food energy intake, lack of physical activity, and genetic susceptibility, though endocrine disorders play a role in a few cases. Obesity is defined by an increase in adipose tissue mass, which is caused by an increase in fat-cell mass [49].

Anti-Inflammatory Activity

Inflammation is an organism's defence response designed to eliminate or limit the spread of harmful agents. There are numerous components of an inflammatory response that can contribute to symptoms and tissue damage [50]. It involves a complex web of intracellular cytokine signals that activate monocytes and/or macrophages, causing them to release a variety of inflammatory mediators such as tumour necrosis factor- (TNF-), interleukin-1 (IL-1), interleukin-6 (IL-6), reactive oxygen species (ROS), prostaglandin E2 (PGE2), and nitric oxide (NO) [51, 52]. MAPKs and the transcription factor NF-kB, in addition, play important roles in mediating extracellular and cellular responses[53]. In mammalian cells, the extracellular signal-regulated kinase (ERK), p38, and c-jun N-terminal kinase are all important MAPKs (JNK) [54]. These mediators are crucial in the regulation of cellular responses to pro-inflammatory molecules, specifically TNF-, IL-1, and IL-63 [55].

TNF- sends cellular signals that activate NF-B, a key factor in regulating gene expression of enzymes and cytokines associated with inflammation, such as iNOS, COX-2, TNF-, IL-1, and IL-6. NF-B regulates the expression of numerous genes involved in immune and inflammatory responses, as well as cellular proliferation, all of which are involved in a variety of diseases [56]. NF-B is a target for the treatment of various inflammatory diseases due to its importance in inflammatory gene expression, and most anti-inflammatory drugs have been shown to inhibit expression of inflammatory cytokines by inhibiting activation via NF-B1,5. It has been demonstrated that selective inhibitors of iNOS, COX-2, TNF-, and IL-1 production and biological activity result in a significant improvement in the development of various inflammatory diseases [57]. GA, like many natural products, is remarkable for its anti-inflammatory properties, which include suppressing proinflammatory cytokines and chemokines such as COX-28,9.

COX-2 overexpression appears to play a role in cancer development by promoting cell division and inhibiting apoptosis. COX-2 is one of many factors associated with tumour progression that is thought to be important in cancer inflammation. Inhibiting COX appears to be directly related to the induction of apoptosis by preventing the release of cytochrome c, caspase activation, and PARP cleavage [58]. Chandramohan et al. demonstrated that GA (25-50 M) reduced COX-2 levels to levels comparable to celecoxib (80-160 mm) [59]. GA is exceptional due to its anti-inflammatory effects by suppressing proinflammatory cytokines, and GA is already known to have anti-proliferative and antitumor activity in lung cancer, prostate cancer, and leukaemia [60, 61]. However, the molecular basis for the activity in inhibiting cancer cell growth and proliferation remains unknown. GA7,12,13 has recently been shown to have potent anti-HAT (histone acetyltransferase) activity [62, 63].

Cardiovascular Activity

Myocardial ischemia is a condition caused by an imbalance in the oxygen supply and demand of the myocardium, with coronary artery atherosclerosis being the most common cause. To reduce the risk of myocardial infarction, hypoperfusion can be treated with various surgical methods and/or pharmacological agents pre-treatment with gallic acid reduces the harmful oxidative consequences of myocardial infarction [63], either by increasing the activity of antioxidant enzymes such as SOD, CAT, GST, and GPx (1) or by increasing the level of non-enzymatic antioxidant agents such as GSH, vitamin C, and vitamin E. (1). All of these activities can help to decrease the harmful effects of free radicals on the integrity and function of myocyte membranes, causing a decrease in serum cardiac biomarkers like cardiac troponin T (cTnT) and creatine kinase-MB (CK-MB) concentrations after an infarction [64].

Conclusion

According to the research presented here, the most crucial pharmacological properties of gallic acid are its antioxidant and anti-inflammatory properties.

Gallic acid is also involved in a variety of signaling pathways that regulate a wide range of biological functions, including pro- and anti-inflammatory pathways, NO signaling pathways, intrinsic and extrinsic apoptosis pathways, and the NF-B signaling pathway. Gallic acid and its derivatives have shown a wide range of beneficial effects in the prevention and/or management of a variety of disorders, and their adequate safety and stability profiles make them viable options for use as dietary supplements.

References

[1] Pengelly A. *The Constituents of Medicinal Plants: An Introduction to the Chemistry and Therapeutics of Herbal Medicine*. 2nd ed. CABI; 2004.

[2] Siah M, Farzaei M, Ashrafi-Kooshk M, Adibi H, Arab S, Rashidi M, Khodarahmi R. Inhibition of guinea pig aldehyde oxidase activity by different flavonoid compounds: an *in vitro* study. *Bioorg Chem*. 2016;64:74–84.

[3] Fernandes F, Salgado H. Gallic acid: review of the methods of determination and quantification. *Crit Rev Anal Chem*. 2016;46:257–265.

[4] Choubey S, Varughese L, Kumar V, Beniwal V. Medicinal importance of gallic acid and its ester derivatives: a patent review. *Pharm Pat Anal*. 2015;4:305–315.

[5] Kasper D, Hauser S, Jameson J, Fauci A, Longo D, Loscalzo J. *Harrison's Principles of Internal Medicine*. 19th ed. Mc Graw Hill; 2015.

[6] Ahmedin J, Freddie B, Melissa M, Jacques F, Elizabeth W, David F. Global cancer statistics. *Cancer J Clin*. 2011;61:69–90.

[7] Giftson J, Jayanthi S, Nalini N. Chemopreventive efficacy of gallic acid, an antioxidant and anticarcinogenic polyphenol, against 1, 2-dimethyl hydrazine induced rat colon carcinogenesis. *Invest New Drugs*. 2010;28:251–259.

[8] Nemec M, Kim H, Marciante A, Barnes R, Talcott S, Mertens-Talcott S. Pyrogallol, an absorbable microbial gallotannins-metabolite and mango polyphenols (Mangifera Indica L) suppress breast cancer ductal carcinoma in situ proliferation *in vitro*. *Food Funct*. 2016;7:3825–3833.

[9] Da Silva S, Chaar J, Yano T. Chemotherapeutic potential of two gallic acid derivative compounds from leaves of Casearia sylvestris Sw (Flacourtiaceae) *Eur J Pharmacol*. 2009;608:76–83.

[10] Huang P, Hseu Y, Lee M, Kumar K, Wu C, Hsu L, Liao JW, Cheng IS, Kuo YT, Huang SY, Yang HL. *In vitro* and *in vivo* activity of gallic acid and Toona sinensis leaf extracts against HL-60 human premyelocytic leukemia. *Food Chem Toxicol*. 2012;50:3489–3497.

[11] Jagan S, Ramakrishnan G, Anandakumar P, Kamaraj S, Devaki T. Antiproliferative potential of gallic acid against diethylnitrosamine-induced rat hepatocellular carcinoma. *Mol Cell Biochem*. 2008;319:51–59.

[12] Chou SF, Luo LJ, Lai JY. Gallic acid grafting effect on delivery performance and antiglaucoma efficacy of antioxidant-functionalized intracameral pilocarpine carriers. *Acta Biomaterialia* 2016; 1(38):116-128.
[13] You Jung KIM. Antimelanogenic and Antioxidant Properties of Gallic Acid. *Biol. Pharm. Bull.* 2007; 30(6):1052-1055.
[14] Ornchuma Naksuriya, Siriporn Okonogi. Comparison and combination effects on antioxidant power of curcumin with gallic acid, ascorbic acid, and xanthone. *Drug Discoveries Therapeutics* 2015; 9(2):136-141.
[15] Monika Bajpai, Anurag Pande, Tewari SK, Prakash Dhan. Phenolic contents and antioxidant activity of some food and medicinal plants. *International Journal of Food Sciences and Nutrition* 2005; 56(4):287-291.
[16] Nathalie SC Gaulejac, Christian Provost, Nicolas Vivas. Comparative Study of Polyphenol Scavenging Activities Assessed by Different Methods. *J. Agric. Food Chem.* 1999; 47:425-431.
[17] Wang K, Zhu X, Zhang K, Zhu L, Zhou F. Investigation of gallic acid induced anticancer effect in human breast carcinoma MCF-7 cells. *J Biochem Mol Toxicol* 2014; 28:387-393.
[18] Usha T, Middha SK, Bhattacharya M, Lokesh P, Goyal AK. Rosmarinic Acid, a New Polyphenol from Baccaurea ramiflora Lour. Leaf: A Probable Compound for Its Anti-Inflammatory Activity. *Antioxidants* (Basel) 2014; 3:830-842.
[19] Locatelli C, Filippin Monteiro FB, Creczynski Pasa TB. Alkyl esters of gallic acid as anticancer agents: a review. *Eur. J. Med. Chem.* 2013; 60:233-239.
[20] Singleton VL. Naturally occurring food toxicants: phenolic substances of plant origin common in foods. *Adv. Food Res.* 1981; 27:149-242.
[21] Priscilla HD, Prince PSM. Cardioprotective effect of gallic acid on cardiac troponin-T, cardiac marker enzymes, lipid peroxidation products and antioxidants in experimentally induced myocardial infarction in Wistar rats. *Chem Biol Interact* 2009; 179(23):118-24.
[22] Borges A, Ferreira C, Saavedra M, Simoes M. Antibacterial activity and mode of action of ferulic and gallic acids against pathogenic bacteria. *Microb Drug Resist.* 2013;19:256–265.
[23] Sanchez-Maldonado A, Schieber A, Ganzle M. Structure–function relationships of the antibacterial activity of phenolic acids and their metabolism by lactic acid bacteria. *J Appl Microbiol.* 2011;111:1176–1184.
[24] Shao D, Li J, Li J, Tang R, Liu L, Shi J, Huang, Q, & Yang, H. Inhibition of gallic acid on the growth and biofilm formation of Escherichia coli and Streptococcus mutans. *J Food Sci.* 2015;80:1299–1305.
[25] Kang M, Oh J, Kang I, Hong S, Choi C. Inhibitory effect of methyl gallate and gallic acid on oral bacteria. *J Microbiol.* 2008;46:744–750.
[26] Teodoro G, Ellepola K, Seneviratne C. potential use of phenolic acids as anti-candida agents-a review. *Front Microbiol.* 2015;6:1420.
[27] Oh B, Jeon E. Synergistic anti-Campylobacter jejuni activity of fluoroquinolone and macrolide antibiotics with phenolic compounds. *Front Microbiol.* 2015;13.
[28] Nohynek L, Alakomi H, Kahkonen M, Heinonen M, Helander I, Oksman-Caldentey K, Puupponen-Pimiä RH. Berry phenolics: antimicrobial properties and

mechanisms of action against severe human pathogens. *Nutr Cancer.* 2006;54:18–32.
[29] Godstime C, Felix O, Augustina O, Christopher O. Mechanisms of antimicrobial actions of phytochemicals against enteric pathogens. *J Pharm Chem Biol Sci.* 2014;2:77–85.
[30] Kubo I, Fujita K, Nihei K, Masuoka N. Non-antibiotic antibacterial activity of dodecyl gallate. *Bioorg Med Chem.* 2003;11:573–580.
[31] Modi M, Goel T, Das T, Malik S, Suri S, Rawat AK, Srivastava SK, Tuli R, Malhotra S, Gupta SK. Ellagic acid & gallic acid from Lagerstroemia speciosa L inhibit HIV-1 infection through inhibition of HIV-1 protease & reverse transcriptase activity. *Indian J Med Res.* 2013;137:540–548.
[32] Kratz J, Andrighetti-Frohner C, Kolling D, Leal P, Cirne-Santos C, Yunes R, Nunes RJ, Trybala E, Bergström T, Frugulhetti IC, Barardi CRM, & Simões CMO. Anti-HSV-1 and anti-HIV-1 activity of gallic acid and pentyl gallate. *Mem Inst Oswaldo Cruz.* 2008;103:437–442.
[33] Zuo G, Li Z, Chen L, Xu X. *In vitro* anti-HCV activities of Saxifraga melanocentra and its related polyphenolic compounds. *Antivir Chem Chemother.* 2005;16:393–398.
[34] Hsu W, Chang S, Lin L, Li C, Richardson C, Lin C, & Lin LT. Limonium sinense and gallic acid suppress hepatitis C virus infection by blocking early viral entry. *Antiviral Res.* 2015;118:139–147.
[35] Lee J, Oh M, Seok J, Kim S, Lee D, Bae G, Bae HI, Bae S, Hong YM, Kwon SO, Lee DH, Song CS, Mun J, Chung M, & Kim K. Antiviral effects of black raspberry (Rubus coreanus) seed and its gallic acid against influenza virus infection. *Viruses.* 2016;6.
[36] Ndjonka D, Abladam E, Djafsia B, Ajonina-Ekoti I, Achukwi M, Liebau E. Anthelmintic activity of phenolic acids from the axlewood tree Anogeissus leiocarpus on the filarial nematode Onchocerca ochengi and drug-resistant strains of the free-living nematode Caenorhabditis elegans. *J Helminthol.* 2014;88:481–488.
[37] Abouelhassan Y, Garrison A, Bai F, Norwood V, Nguyen M, Jin S, Huigens RW. A Phytochemical-halogenated quinoline combination therapy strategy for the treatment of pathogenic bacteria. *Chem Med Chem.* 2015;10:1157–1162.
[38] Shibata H, Kondo K, Katsuyama R, Kawazoe K, Sato Y, Murakami K, Takaishi Y, Arakaki N, Higuti T. Alkyl gallates, intensifiers of β-lactam susceptibility in methicillin-resistant Staphylococcus aureus. *Antimicrob Agents Chemother.* 2005; 49:549–555.
[39] Manvar D, Mishra M, Kumar S, Pandey VN (2012) Identification and evaluation of anti hepatitis C virus phytochemicals from Eclipta alba. *J Ethnopharmacol* 144: 545-554.
[40] Mehrab MM, Sendi H, Steuerwald N, Ghosh S, Schrum LW, Bonkovsky HL. (2011) Legalon-SIL downregulates HCV core and NS5A in human hepatocytes expressing full-length HCV. *World J Gastroenterol* 17: 1694-1700.
[41] Micha R, Kalantarian S, Wirojratana P, Byers T, Danaei G, Elmadfa I, Ding E, Giovannucci E, Powles J, Smith-Warne, S, Ezzati M, Mozaffarian D, & on behalf of the Global Burden of Diseases, Nutrition and Chronic Disease Expert Group.

(2012) Estimating the global and regional burden of suboptimal nutrition on chronic disease: Methods and inputs to the analysis. *Eur J Clin Nutr* 66: 119-129.

[42] Nair HB, Sung B, Yadav VR, Kannappan R, Chaturvedi MM, & Aggarwal B. (2010) Delivery of antiinflammatory nutraceuticals by nanoparticles for the prevention and treatment of cancer. *Biochem Pharmacol* 80: 1833-1843.

[43] Latha RCR, Daisy P (2011) Insulin-secretagogue, antihyperlipidemic and other protective effects of gallic acid isolated from Terminalia bellerica Roxb. In streptozotocin-induced diabetic rats. *Chem Biol Interact* 189: 112-118.

[44] Chhhikara N., Kaur R., Jaglan S, Sharma P, Gat Y, Panghal A. Bioactive compounds and pharmacological and food applicationsof Syzygium cumini-A review. *Food Funct*. 2018,9, 6096–6115. [CrossRef].

[45] Naveed M, BiBi J, Kamboh AA, Suheryani, I, Kakar I, Fazlani SA, FangFang X, Kalhoro SA, Yunjuan L, Kakar MU, Abd El-Hack ME, Noreldin AE, Zhixiang S, LiXia C, & XiaoHui Z. Pharmacological values and therapeutic properties of black tea (Camellia sinensis): A comprehensive overview. *Biomed. Pharmacother*. 2018, 100, 521–531.

[46] Li S, Lo, CY, Pan MH, Lai CS, Ho CT. Black tea: Chemical analysis and stability. *Food Funct*. 2013,4,10–18.

[47] Pastoriza S, Mesias M, Cabrera C, Rufian-Henares JA. Healthy properties of green and white teas: Anupdate. *Food Funct*. 2017,8,2650–2662.

[48] Oliveira MR, Nabavi, SF, Daglia M, Rastrelli L, Nabavi SM. Epigallocatechin gallate and mitochondria-A story of life anddeath. *Pharmacol. Res*. 2016,104, 70–85.

[49] Arfan M, Amin H, Khan N, Khan I, Saeed M, Khan MA, Fazal-ur-Rehman. Analgesic and anti-inflammatory activities of 11-O-galloylbergenin. *J Ethnopharmacol* 2010, 131, 502-4.

[50] Na HJ, Lee G, Oh HY, Jeon KS, Kwon HJ, Ha KS, Lee H, Kwon YG, Kim YM. 4-O-Methylgallic acid suppresses inflammation-associated gene expression by inhibition of redox-based NF-kappaB activation. *Int Immunopharmacol* 2006, 6, 1597-608.

[51] Chang YH, Lee ST, Lin WW. Effects of cannabinoids on LPS-stimulated inflammatory mediator release from macrophages: involvement of eicosanoids. *J Cell Biochem* 2001, 81, 715-23.

[52] Azzolina A, Bongiovanni A, Lampiasi N. Substance P induces TNF-alpha and IL-6 production through NF kappa B in peritoneal mast cells. *Biochim Biophys Acta* 2003, 1643, 75-83.

[53] Beyaert R, Cuenda A, Vanden Berghe W, Plaisance S, Lee JC, Haegeman G, Cohen P, Fiers W. The p38/RK mitogen-activated protein kinase pathway regulates interleukin-6 synthesis response to tumor necrosis factor. *EMBO J* 1996, 15, 1914-23.

[54] Lee G, Na HJ, Namkoong S, Jeong Kwon H, Han S, Ha KS, Kwon YG, Lee H, Kim YM. 4-O-methylgallic acid down-regulates endothelial adhesion molecule expression by inhibiting NF-kappaB-DNA-binding activity. *Eur J Pharmacol* 2006, 551, 143-51.

[55] Ghosh S, May MJ, Kopp EB. NF-kappa B and Rel proteins: evolutionarily conserved mediators of immune responses. *Annu Rev Immunol* 1998, 16, 225-60.
[56] Choi KC, Lee YH, Jung MG, Kwon SH, Kim MJ, Jun WJ, Lee J, Lee JM,; Yoon HG. Gallic acid suppresses lipopolysaccharide-induced nuclear factor-kappaB signaling by preventing RelA acetylation in A549 lung cancer cells. *Mol Cancer Res* 2009, 7, 2011-21.
[57] Jung HJ, Kim SJ, Jeon WK, Kim BC, Ahn K, Kim K, Kim YM, Park EH, Lim CJ. Anti-inflammatory activity of n-propyl gallate through down-regulation of NF-κB and JNK pathways. *Inflammation* 2011, 34, 352-61.
[58] Yoon CH, Chung SJ, Lee SW, Park YB, Lee SK, Park MC. Gallic acid, a natural polyphenolic acid, induces apoptosis and inhibits proinflammatory gene expressions in rheumatoid arthritis fibroblast-like synoviocytes. *Joint Bone Spine* 2012.
[59] Chandramohan Reddy, T, Bharat Reddy D, Aparna A, Arunasree KM, Gupta G, Achari C, Reddy GV, Lakshmipathi V, Subramanyam A, Reddanna P. Anti-leukemic effects of gallic acid on human leukemia K562 cells: downregulation of COX-2, inhibition of BCR/ABL kinase and NF-κB inactivation. *Toxicol In Vitro* 2012, 26, 396-405.
[60] Wang MT, Honn KV, Nie D. Cyclooxygenases, prostanoids, and tumor progression. *Cancer Metastasis Rev* 2007, 26, 525-34.
[61] Jagan S, Ramakrishnan G, Anandakumar P, Kamaraj S, Devaki T. Antiproliferative potential of gallic acid against diethylnitrosamine-induced rat hepatocellular carcinoma. *Mol Cell Biochem* 2008, 319, 51-9.
[62] Madlener S, Illmer C, Horvath Z, Saiko P, Losert A, Herbacek I, Grusch M, Elford HL, Krupitza G, Bernhaus A, Fritzer-Szekeres M, Szekeres T. Gallic acid inhibits ribonucleotide reductase and cyclooxygenases in human HL-60 promyelocytic leukemia cells. *Cancer Lett* 2007, 245, 156-62.
[63] Priscilla D, Prince P. Cardioprotective effect of gallic acid on cardiac troponin-T, cardiac marker enzymes, lipid peroxidation products and antioxidants in experimentally induced myocardial infarction in wistar rats. *Chem Biol Interact*.
[64] Kasper D, Hauser S, Jameson J, Fauci A, Longo D, Loscalzo J. *Harrison's Principles of Internal Medicine*. 19th ed. Mc Graw Hill; 2015.

Chapter 6

Gallic Acid: A Potential Antidiabetic Agent

Suraj Tarihalkar[*]
Venkatesh Kumbhar[†]
Poornima Sankpal[‡]
and Sachinkumar Patil[§]

Ashokrao Mane College of Pharmacy,
Peth-Vadgaon, Kolhapur, Maharashtra, India

Abstract

Diabetes mellitus (DM), which increases morbidity and mortality, has emerged as a global health issue. Gallic acid is a phenolic molecule with antidiabetic properties. The creation and activation of oxidative stress were originally linked to the destructive course of diabetes mellitus (DM). Inflammation and oxidative stress have both been proven to have a significant role in the pathological development of diabetes mellitus and its associated consequences.

Gallic acid (GA) was studied for its potential to treat diabetes mellitus using natural antioxidants. Gallic acid has consistently shown strong anti-inflammatory and anti-oxidative effects on metabolic illnesses. Gallic acid is commonly available in herbal forms and edible plants. Gallic acid is a potent compound which shows anti-diabetic effects. A summarization of derivatives is required.

The objective of this chapter is to highlight the latest theories and findings in the fields of oxidative stress and diabetes mellitus. For

---

[*] Corresponding Author's Email: tarihalkarsuraj262@gmail.com.
[†] Corresponding Author's Email: venketeshkumbhar58649@gmail.com.
[‡] Corresponding Author's Email: poournima6@gmail.com.
[§] Corresponding Author's Email: sachinpatil.krd@gmail.com.

In: The Chemistry of Gallic Acid and Its Role in Health and Disease
Editor: Jeff C. Murdoch
ISBN: 979-8-88697-672-4
© 2023 Nova Science Publishers, Inc.

diabetes mellitus and its consequences, Gallic acid functions as an antiglycemic drug.

Keywords: gallic acid, diabetes mellitus, oxidative stress, anti-diabetic effects

Introduction

One of the most significant public health issues, diabetes mellitus (DM), is a chronic metabolic illness that affects millions of people worldwide [1] (Mehrab-Mohseni, Marjan. et al. 2011). Hyperglycaemia, which is a result of deficiencies in insulin production and secretion, is the main characteristic of this condition. Polydipsia, polyphagia, polyuria, weight loss, weariness, and vision loss have all been listed as symptoms of DM. These signs may be negligible or even nonexistent [2] (Manvar, Dinesh, Mahesh Mishra. Et al. 2012).

According to research by De Oliveira et al., gallic acid has an impact on histology and biochemical markers as well as oxidative stress indicators. Since gallic acid lowers TG, TC, and LDL levels, it promotes a shift in the lipid profile of the control animals. In addition to lowering the amounts of free radicals and lipid peroxidation, it also increases the protection provided by non-enzymatic and enzymatic cancer prevention agents in these tissues. Their findings demonstrate the potential use of gallic acid in the management of hepatic and renal complications associated with diabetes mellitus and increase the chance that it will find use in a similar manner to hypoglycaemic medicines. According to a study, conducted by Efthimios et al. the natural flavonoids gallic acid and ellagic acid can be used as antidiabetic medications. Nair et al. examined the interaction between glycogen phosphorylase (GP), a crucial contender for maintaining glucose homeostasis, and gallic acid and its dimer, ellagic acid. Both bioactive substances work as non-competitive inhibitors of the allosteric activator AMP and competitive inhibitors of glucose-1-phosphate. Ellagic acid may therefore be used as an anti-hyperglycemic drug.

This dimer has a more potent inhibitory impact than gallic acid. Both gallic acid and ellagic acid are bioactive compounds and act as potent inhibitors of glucose-1-phosphate and non-competitive inhibitors of the conformational stimulator AMP. Ellagic acid can consequently be used as a medication to treat hyperglycemia [3] (American Diabetes Association. et al. 2014). There are already 425 million cases of hyperglycemia worldwide, and by 2045, that number is projected to rise by 48% [4] (Federation, International Diabetes. et al. 2017). Diabetes mellitus (DM) is a long-term metabolic condition brought on by insulin failure, which results in hyperglycemia [5] (Mishra, Nidhi. et al. 2013). It caused many organs to fail, malfunction, and sustain long-term harm [6] (Wild, Sarah. et al. 2004).

One treatment for post meal hyperglycemia is to restrict glucose absorption by blocking enzymes that hydrolyze carbohydrates, such as glucosidase. When the glucosidase enzyme is not present, blood glucose levels might revert to normal [7] (Bosenberg L. et al. 2008). Some naturally occurring phenolic compounds and polysaccharides (such as those found in tea, grapefruit, and strawberries) have recently been discovered to inhibit the glucosidase enzyme, suggesting that they may have potential as natural therapeutic agents for the treatment of diabetes mellitus [8] (Chen, Haixia. et al. 2020). The increased production of pro-oxidants during diabetes, which results in oxidative damage to biological molecules, cells, tissues, and organs, oxidative stress is indeed a major mediator in the initiation and progression of diabetic problems [9] (Pourghassem-Gargari, Bahram. et al. 2011). The redox system perturbation caused by noticeably increased reactive oxygen compound generation leads to oxidative stress. It was determined that DM is an oxidation state caused by elevated ROS formation. The OS pathway may be engaged to cause hyperglycemia-induced damage and insulin resistance inflammation in host organs even as levels of ROS are high in a hyperglycemia environment.

Through the traditional position of ROS as second messengers in cell signaling, oxidative stress actively promotes cellular injury and death by triggering oxidative damage in cellular constituents, including DNA degradation and the peroxidation of proteins and lipids. Oxidative stress may damage the insulin-producing pancreatic cell directly & reduce insulin sensitivity, which would then cause DM to develop.

Mineral supplements may be a prospective target for the creation of effective anti-diabetic drugs, according to recent developments in the field of drug discovery [10, 11] (Mwiti Kibiti, Cromwell. et al. 2015) (Kimball, Samantha M. et al. 2017). This is because pharmacological studies have supported their modulatory effects on glucose and lipid metabolism in diabetes, obesity, and associated metabolic diseases, while they serve as essential coenzymes and cofactors for metabolic processes that keep normal glucose, lipid, and protein metabolism [12] (Mooradian, Arshag D. et al. 1994). Particularly, it has recently been observed that Zn (II) exhibits comparatively significant insulin-mimetic characteristics while being less hazardous than other metal elements [13, 14] (Adachi, Yusuke. et al. 2004) (Chukwuma, Chika Ifeanyi. et al. 2020). Zn (II) increases both lipogenesis and glucose transport in adipocytes [15, 16] (Ezaki, Osamu. et al. 1989) (Shisheva, Assia. et al. 1992). Zn II altered insulin signalling in skeletal muscle cells from humans and animals, improving glucose oxidation and glycaemic control [17] (Norouzi, Shaghayegh. et al. 2018). It might thus represent a possible target for the creation of a diabetes treatment drug.

Gallic Acid

One of the drug-like properties of gallic acid, a phenolic compound, is whether it acts as an anti-diabetic. The use of gallic acid as a treatment for diabetes mellitus (DM) is unsatisfactory since it is unstable at high temperatures, in the presence of oxygen, or in the presence of light. It is also suspected that less of the treatment is absorbed into the body because if the large particle size and poor solubility during the absorption process [18] (Jacques, Andresa Carolina. et al. 2010).

It is encouraging to perform the encapsulation of an active compound in nanoparticles to alter its solubility and bioavailability. The encapsulation technique is used to shield the bioactive chemicals from the effects of oxygen, heat, and light that might impair the stability of the compound by encapsulating them in a secondary ingredient (polysaccharide, protein, or lipid) [19] (Pillai, Dipin S. et al. 2012).

Diabetes Mellitus and Related Complications: Oxidative Stress

Hyperglycemia and Oxidative Stress

Hyperglycaemia, by causing glycation and glycoxidation of cellular proteins, has the potential to impair organ metabolic activity and cause relevant cell or tissue injury at the sites of diabetic complications. Oxidative Stress pathways may be a critical link between tissue damage and changes in glucose homeostasis. Hyperglycaemia has been proposed as the primary cause of chronic or prolonged oxidative stress in diabetes-related tissue injury. Another significant imbalance between the production of ROS and the antioxidant defence in organs might result from hyperglycaemic circumstances. The oxidation of lipids or glucose and the decreased activity of antioxidant enzymes like SOD, GPX, and CAT, which are all implicated in the antioxidant defence of tissues in diabetes circumstances, are the hyperglycaemia-related sources of ROS. Homeostasis of cellular redox may be associated with increased activation of redox-sensitive genes [20], (Muriach, María. et al. 2014) the significance of which will be highlighted by the following list:

1. Less GSH is produced from glutathione disulphide due to NADPH depletion via the elevated glucose-induced polyol pathway, which in turn contributes to ROS production and OS.
2. Hyperglycaemia-mediated protein kinase C (PKC) raises the content of diacylglycerol and increases NADPH oxidase (NOX) activity.
3. Through NADPH oxidase and the mitochondrial field, the increased interaction between advanced glycation end products (AGEs) and RAGEs promotes production [21] (Cepas, Vanesa. et al. 2020).

Insulin Resistance & Oxidative Stress

Insulin resistance is a major contributor to the development of metabolic syndrome and diabetes mellitus. Depletion of hydrogen peroxide can improve insulin resistance [22], (Onyango, Arnold N. 2018) and a high level of ROS detected inside the hepatocytes of obese mice can induce insulin resistance. Multiple stress-responsive pathways are induced as insulin resistance develops pathologically, resulting in cell malfunction, autophagy, apoptosis, or death in pancreatic cells that are essential for glucose and insulin control. Insulin resistance causes the expression of regulatory cytokines to increase, stress-responsive pathways like JNK to become active, and linked proteins to sustain direct oxidative stress. This OS can also reduce insulin sensitivity in response to insulin-related signaling pathways. For instance, the increased ROS may prevent the PI3K-Akt signaling pathway from activating and dephosphorylating the insulin receptors, which further reduces the translocation of glucose transporter 4 (GLUT4), which is required for glucose absorption.

Gallic Acid: Natural Bioactive Metabolite

Gallic acid is frequently produced by Hydrolyzing polyphenol tannic acid [23] (Andrade, Priscilla Macedo Lima. et al. 2018) or tannins (Gallo tannins and ellagitannins). Additionally, GA could be produced through biological, enzymatic, acidic, and alkaline processes. Tannic acid might be hydrolysed with an inducible hydrolase of Enterobacter species [24] (Sharma, Kanti Prakash. et al. 2017) to produce gallic acid. There are numerous edible plants, such as fruits, berries, and nuts have substantial yields of tannic acid (persimmons, apples, grapes, and almonds). In cranberries, blueberries, raspberries, and walnuts, ellagic acid (EA), a condensed dimer of GA, can be discovered either in the bonded formation (ellagitannins) or in combination with hexahydroxy diphenic acid. EA possesses several pharmacological properties, including antioxidant and antihyperglycemic properties, as well as the ability to control apoptosis-inducing activities, making it easier to treat a variety of chronic diseases in humans [25] (Shakeri, Abolfazl. et al. 2018).

Gallic Acid: Natural Antioxidant

Gallic Acid in Plants

A frequent secondary metabolite abundantly found in many different food plants, including gall nuts, tea leaves, oak bark, blueberries, grape seeds, rose flowers, sumac, witch hazel, the fruit of *Syzygium cumini,* etc., is gallic acid is phenolic compound with the molecular formula $C_6H_2(OH)_3COOH$ [26] (Chhikara, Navnidhi. et al. 2018).

These edible crops with high gallic acid content may have a high nutritive value due to gallic acid and its derivatives' biological functions. Tea, for example, was discovered to have antioxidant and cholesterol-lowering properties after extensive research [23] (Andrade, Priscilla Macedo Lima. et al. 2018).

Gallic Acid as a Natural Bioactive Metabolite

Hydrolyzing polyphenol tannic acid and tannins (such as gallo and ellagitannins) yields GA [27] (Naveed, Muhammad. et al. 2018). Additionally, GA could be produced through biological, enzymatic, acidic, and alkaline processes. Tannic acid might be hydrolysed by an inducible hydrolase of Enterobacter species to produce GA [24] (Sharma, Kanti Prakash. et al. 2017) several plants that are edible, such as fruits, berries, and nuts, have substantial yields of tannic acid (persimmons, apples, grapes, and almonds). The compressed dimer of GA known as ellagic acid (EA), which is present in cranberries, blueberries, raspberries, and walnuts, can be found either in the bonded formation (ellagitannins) or in combination with hexahydroxy diphenic acid. EA demonstrates a wide range of pharmacological qualities, such as antioxidants and antihyperglycemic actions, additional controlling apoptosis-inducing actions, which are helpful for the treatment of numerous chronic human disorders [28] (Sharma, Kanti Prakash. et al. 2017).

Tannins in humans are hydrolysed in the digestive tract before being absorbed or broken down by microorganisms. Inside the colon, enzymatic and non-enzymatic hydrolysis of gallo tannins may release free gallic acid, and bacterial decarboxylation of Gallo tannin may result in the production of pyrogallol and gallic acid [29] (Shakeri, Abolfazl, Mohammad R. et al. 2019). Diabetes patients, classified as type 1 and type 2 DM, are invariably associated with the long-term metabolic syndrome. The main contributor to the onset of diabetes mellitus (DM) is the ensuing hyperglycaemia or insulin resistance. The development of antioxidant therapy for DM and its associated problems has increased interest in natural antioxidants with a protective effect against OS. Through the natural role of ROS as second messengers in cell signaling, OS induces oxidative damage in cellular constituents, including DNA degradation and the peroxidation of proteins and lipids, resulting in cellular injury and death. OS may directly damage the pancreatic cells that produce insulin and reduce insulin sensitivity, which will lead to DM.

Gallic acid has shown excellent antioxidant and free radical scavenging properties that could restore the metabolism to its regular state. Gallic acid consumption improved serum measurements in obese rats while lowering oxidative stress and promoting tumour-associated glycoprotein (TAG), phospholipid, total cholesterol, LDL-cholesterol, insulin, and leptin levels.

Gallic Acid as a Versatile Antioxidant Agent

The trihydroxy benzoic acid gallic acid, also known as 3,4,5-trihydroxybenzoic acid, has the formula $C_6H_2(OH)_3CO_2H$. It is under the phenolic acid category. Gallnuts, sumac, witch hazel, tea leaves, oak bark, and other plants also contain it. Although samples are often dark due to partial oxidation, it is a white solid. "Gallates" are the name for gallic acid salts & esters.

Oxidative stress is a major contributor to the development of various degenerative illnesses, including cancer, atherosclerosis, cardiovascular disease, ageing, and inflammatory diseases. Gallic acid (3,4,5-trihydroxybenzoic acid), a low molecular weight triphenolic molecule, has emerged as a powerful antioxidant and apoptosis inducer.

Structure Modification of Gallic Acid

GA's structure has been altered to change its polarity, making it more water-soluble and absorbable through the enterocyte membrane, and thus increasing its nutritional value [30] (Abdou, Ebtsam Mohmmed. et al. 2018). Nano formulations might improve the biodegradability and bioavailability of these substances. The bioavailability of GA is increased, as is its lipophilicity, by complexing with hydrogenated soy phosphatidylcholine (HSPC). Hepatotoxicity caused by CCL4 was much more resistant to antioxidant degradation with the GA-HSPC complex nanoformulation [31] (Bhattacharyya, Sauvik. et al. 2013). Gallic acid is a member of the group of chemical compounds known as gallic acids. It is sometimes referred to as gallate, acid, or gallic. These are organic substances that have a moiety of 3,4,5-trihydroxybenzoic acid. A very weakly basic (basically neutral) chemical is gallic acid (based on its pKa). All living things, from bacteria to people, contain gallic acid. Gallic acid is typically most plentiful outside of the human body in a few foods, such as mango, pomegranates, and cloves, and is least abundant in red raspberries, cumin, and turnips. Several other foods, including common walnuts, tarragon, ginger, fruit juices, and corn, have also been found to contain gallic acid; however, the amount has not been defined.

Gallic Acid for Diabetic Therapy

In certain pathological conditions, GA and its derivatives can modulate oxidative stress, apoptosis, or inflammation since they are powerful antioxidants and free radical scavengers. GA has antihyperglycemic potential due to its antioxidant and anti-inflammatory properties [32] (D'Souza, Jason Jerome. et al. 2014). In adipose tissues, GA from E. officinalis fruit juice enhanced insulin sensitivity and glucose homeostasis. Mechanistically, combining PPAR- and C/EBP activation increased GLUT4 translocation in adipocytes.
Furthermore, evidence of simultaneous AMPK and Akt stimulation by *E. officinalis* fruit juice [33] (Kahkeshani, Niloofar. et al. 2019) suggests that GA may improve insulin sensitivity by controlling the AMPK and Akt signalling pathways. According to the results, PPAR, Akt, and AMPK stimulation helped GA's antidiabetic effects. Furthermore, GA's anti-diabetic effects may be mediated by controlling the expression of adipocytokines and TNF-. By

preventing caspase-9-related cell death, GA enhanced the ability of cells to operate. The evolution of several diabetic with oxidative damage problems, including nephropathy, was attributed to AGE/ALE end products. Advanced glycation inhibition was induced by glyoxal, which also caused ROS production, membrane lysis, lysosomal membrane leakage, mitochondrial membrane potential collapse, and lipid peroxidation. GA may counteract these effects. Patients with diabetes frequently experience persistently poor treatment of wounds, which promotes viral infections and demands amputation.

Gallic acid may hasten these same human fibroblasts and keratinocyte migrating in addition to activating particles that are known to promote wound healing, including extracellular signal-regulated kinases (ESRK), c-Jun N-terminal kinases (JNK), and focal adhesion kinases (FAK). This mechanism may explain GA's beneficial role in the treatment of wounds caused by diabetes. The antioxidative potential of this polyphenol molecule for diabetic complications altered several anti-diabetic signaling pathways, and we listed some prospective anti-diabetic plants that are nutritious and therapeutic indicated gallic acid as a key component. While oxidative stress-induced liver & kidney damage occurs in the diabetic condition, gallic acid may reduce diabetic nephropathy (DN), neuropathy, and cardiac problems.

Diabetic Nephropathy

Through the scavenging of free radicals which are reactive and enhancing capability for intracellular antioxidants, gallic acid may reduce the damage that SA causes in the kidneys and liver. By reducing oxidative stress and microRNAs linked to gallic acid reduced renal fibrosis, endoplasmic reticulum stress and methylglyoxal-induced DN. In type I diabetic rats, gallic acid therapy significantly reduced levels of albumin, blood urea nitrogen, and serum creatinine and increasing creatinine levels & decreasing TGF-1 level.

When compared to certain other polyphenols in food, gallic acid (GA), a polyphenol component spotted in vegetables, red wine, tea, as well as other types of food, does have a comparatively straightforward structure. Even though much research has indicated that GA aids in the therapy for metabolic illnesses, previous research has primarily used conventional biomarkers and histopathological sections to conduct analysis, leading to a limited knowledge of the roles played by GA. Therefore, it is important to revaluate the roles and processes of GA in treating metabolic disorders.

In this study, the utility of GA to treat diabetes and non-alcoholic fatty liver disease brought on by high-fat diet (HFD) and streptozotocin was explored using a pharmacodynamic mouse model (STZ). Additionally, metabolite alterations in mouse blood, urine, liver, or muscle tissues were examined using a 1H nuclear magnetic resonance-based metabolomics study, confirming the therapeutic impact and probable mechanisms of GA in treating diabetes and NAFLD. The findings showed that HFD and STZ caused serious metabolic problems in diabetic and NAFLD mice, intestinal microbiota changes and disturbances in the metabolism of glucose, lipids, amino acids, purines, and pyrimidines.

GA therapy reduced the mice's elevated blood glucose levels, slowed the development of NAFLD, and partially restored the mice's disrupted metabolic pathways. This study was the first to show that increased oxidation and ketogenesis are linked to the GA mechanism in decreasing lipid build-up. This discovery is consistent with those of earlier pharmacodynamics research and makes it easier to pinpoint new ways that GA can treat metabolic illnesses.

Gallic Acid Derivatives and Gallic Acid

Fruits and plants contain large amounts of gallic acid and structurally similar chemicals. There is a presence of gallic acid and its catechin derivatives in phenolic components contains both green and black tea. Studies utilising these compounds have found them to possess several therapeutic possibilities, particularly anti-cancer, and anti-microbial properties. Gallic acid, a naturally occurring phenic acid derived from ingestible plants, has been used in nutraceutical goods as an anti-inflammatory & immune system booster. The multiple health benefits of GA can largely be attributed to its capacity for radical scavenging, which helps in preventing or treating oxidative stress, which plays a significant role in diabetes complications and DM.

This study focused on how OS affected pathological development of DM and provided a summary of the most recent research on GA's ability to regulate OS in DM and its associated problems. The phenolic acids of plant metabolites recognised as gallic acid derivatives, including tetradecyl gallate, lauryl gallate, propyl gallate, octyl gallate and hexadecyl gallate, were widely found across the plant kingdom [34] (Fernandes, Felipe Hugo Alencar. et al. 2016). The hydroxy groups are located at locations 3, 4, and 5 in the trihydroxy benzoic acid known as gallic acid. Different lengths of carbocyanines connected to a carboxyl group are present in its derivatives [35] (Brewer, M.

S. et al. 2011). There are numerous uses for gallic and its derivatives in the food, cosmetic, printing, dyeing, & pharmaceutical sectors [36, 37] (Saeed, Naima. et al. 2012) (Kosuru, Rekha Yamini. et al. 2018). Gallic acid can be used as an addition to stop the astringency and deterioration of oils and fats in a variety of foods, including condiments, candies, beverages, and baked and fried foods [38, 39] (Su, Tzu-Rong, Jen-Jie Lin. et al. 2013) (Sawa, Tomohiro. et al. 1999). Gallic acid is used as a critical gradient in many cosmetics because it was discovered that it could prevent melanogenesis, thereby reducing pigmentation and shielding cells from UV-B or ionizing radiation [40, 41] (Shao, Dongyan, Jing Li. et al. 2015) (Sorrentino, Elena. et al. 2018).

In addition to those, there are several publications on the medical benefits with gallic acid, including its anti-inflammatory, anti-bacterial, and anti-allergy, and oxidative stress properties. *Listeria monocytogenes, Chromobacterium violaceum, Staphylococcus aureus, Pseudomonas aeruginosa, Klebsiella pneumoniae, Streptococcus mutans* and *Escherichia coli* were only a few of the pathogens that the gallic acid showed antibacterial properties against [42, 43] (Fu, Rao, Yuting Zhang. et al. 2015) (Hyun, Ki Hyeob. et al. 2019). Gallic acid has reportedly been found to be a successful treatment for allergic contact dermatitis if combined with other phenolic chemicals [44] (Radan, Maryam. et al. 2019). Gallic acid can block the production of mediators that promote inflammation, such as interleukin (IL)-2, IL-4, IL-5, IL-13, and IL-33, tumour necrosis factor (TNF), cyclooxygenase-2 (COX-2), interferon (IFN), and nuclear factor b [45, 46] (Wang, Xinhua. et al. 2018) (Dludla, Phiwayinkosi V. et al. 2018). Gallic acid is an important choice for use as therapeutic agents or nutritional supplements because it is a powerful antioxidant that can protect human cells from both acute and long-term oxidative stress [47] (Variya, Bhavesh C. et al. 2020).

Conclusion

Hyperglycaemia, glucose intolerance, inflammation, and oxidative stress are typically thought of as the main causal components connected to the rapidly deteriorating course of diabetes-related disturbances. It has been discovered that the physiological and pathological advancement of diabetes and the problems associated with different complex mechanisms. Therefore, ameliorating inflammation and oxidative stress may be another basic modification performed to enhance cellular performance in the diabetic condition, in addition to exerting control insulin levels and blood glucose concentration. There are presently very few efficient antioxidant therapies available to help diabetic individuals with their metabolic syndrome. Consuming foods high in polyphenols has been linked to several multitarget antioxidative activities thus far. For example, GA, the chemical compound that all polyphenols share, has demonstrated encouraging outcomes in the treatment of DM and its comorbid problems.

Through the activation of numerous efficient pathways, including production of radicals, the GPX/GSH system, NO/iNOS synthesis, AGE/ALE, JNK/ERK, GLUT4, and Nrf 2 pathways, a few organic ingredients, GA, among others, have been demonstrated to enhance inflammation, hyperglycaemia, insulin, and oxidative stress in diabetic complications. Preclinical information compiled in this research undoubtedly supported the use of GA or its derivatives for health benefits in reducing diabetes-related problems. To find out if it has therapeutic value as an adjunct therapy, research examining its combined use with current glucose-lowering medicines can be added to this. Because edible and herbal plants contain large amounts of GA, this provides proof of the plant's ability to treat diabetes, associated plants and outlines the important anti-inflammatory and antioxidative role played by GA in mediating antidiabetic activity, offering a potential treatment for diabetes (Mehrab-Mohseni, Marjan, et al. 2011).

References

[1] Mehrab-Mohseni, Marjan, Hossein Sendi, Nury Steuerwald, Sriparna Ghosh, Laura W. Schrum, and Herbert L. Bonkovsky. "Legalon-SIL downregulates HCV core and NS5A in human hepatocytes expressing full-length HCV." *World Journal of Gastroenterology: WJG* 17, no. 13 (2011): 1694.

[2] Manvar, Dinesh, Mahesh Mishra, Suriender Kumar, and Virendra N. Pandey. "Identification and evaluation of anti-hepatitis C virus phytochemicals from Eclipta alba." *Journal of Ethnopharmacology* 144, no. 3 (2012): 545-554.

[3] American Diabetes Association. "Diagnosis and classification of diabetes mellitus." *Diabetes Care* 37, no. Supplement_1 (2014): S81-S90.

[4] Federation, International Diabetes. "IDF diabetes atlas 8th edition." *International Diabetes Federation* (2017): 905-911.

[5] Mishra, Nidhi. "Hematological and hypoglycemic potential Anethum graveolens seeds extract in normal and diabetic Swiss albino mice." *Veterinary World* 6, no. 8 (2013).

[6] Wild, Sarah, Gojka Roglic, Anders Green, Richard Sicree, and Hilary King. "Global prevalence of diabetes estimates for the year 2000 and projections for 2030." *Diabetes Care* 27, no. 5 (2004): 1047-1053.

[7] Bosenberg, L. H., and Danie Gerhardus Van Zyl. "The mechanism of action of oral antidiabetic drugs: A review of recent literature." *Journal of Endocrinology, Metabolism and Diabetes in South Africa* 13, no. 3 (2008): 80-88.

[8] Chen, Haixia, Yanan Jia, and Qingwen Guo. "Polysaccharides and polysaccharide complexes as potential sources of antidiabetic compounds: A review." *Studies in Natural Products Chemistry* 67 (2020): 199-220.

[9] Pourghassem-Gargari, Bahram, Somayeh Abedini, Hossein Babaei, Akbar Aliasgarzadeh, and Parvin Pourabdollahi. "Effect of supplementation with grape seed (Vitis vinifera) extract on antioxidant status and lipid peroxidation in patient with type II diabetes." *J Med Plants Res* 5, no. 10 (2011): 2029e34.

[10] Mwiti Kibiti, Cromwell, and Anthony Jide Afolayan. "The biochemical role of macro and micro-minerals in the management of diabetes mellitus and its associated complications: a review." *International Journal for Vitamin and Nutrition Research* 85, no. 1-2 (2015): 88-103.

[11] Kimball, Samantha M., JC Herbert Emery, and Richard Z. Lewanczuk. "Effect of a vitamin and mineral supplementation on glycemic status: results from a community-based program." *Journal of Clinical & Translational Endocrinology* 10 (2017): 28-35.

[12] Mooradian, Arshag D., Mark Failla, Byron Hoogwerf, Melinda Maryniuk, and Judith Wylie-Rosett. "Selected vitamins and minerals in diabetes." *Diabetes Care* 17, no. 5 (1994): 464-479.

[13] Adachi, Yusuke, Jiro Yoshida, Yukihiro Kodera, Akira Kato, Yutaka Yoshikawa, Yoshitane Kojima, and Hiromu Sakurai. "A new insulin-mimetic bis (allixinato) zinc (II) complex: structure–activity relationship of zinc (II) complexes." *JBIC Journal of Biological Inorganic Chemistry* 9, no. 7 (2004): 885-893.

[14] Chukwuma, Chika Ifeanyi, Samson S. Mashele, Kenneth C. Eze, Godfrey R. Matowane, Shahidul Md Islam, Susanna L. Bonnet, Anwar EM Noreljaleel, and Limpho M. Ramorobi. "A comprehensive review on zinc (II) complexes as anti-diabetic agents: The advances, scientific gaps and prospects." *Pharmacological Research* 155 (2020): 104744.

[15] Ezaki, Osamu. "IIb group metal ions (Zn^{2+}, Cd^{2+}, Hg^{2+}) stimulate glucose transport activity by post-insulin receptor kinase mechanism in rat adipocytes." *Journal of Biological Chemistry* 264, no. 27 (1989): 16118-16122.

[16] Shisheva, Assia, Dov Gefel, and Yoram Shechter. "Insulinlike effects of zinc ion *in vitro* and *in vivo*: preferential effects on desensitized adipocytes and induction of normoglycemia in streptozocin-induced rats." *Diabetes* 41, no. 8 (1992): 982-988.

[17] Norouzi, Shaghayegh, John Adulcikas, Sukhwinder Singh Sohal, and Stephen Myers. "Zinc stimulates glucose oxidation and glycemic control by modulating the insulin signaling pathway in human and mouse skeletal muscle cell lines." *PLoS One* 13, no. 1 (2018): e0191727.

[18] Jacques, Andresa Carolina, Paula Becker Pertuzatti, Milene Teixeira Barcia, Rui Carlos Zambiazi, and Josiane Freitas Chim. "Estabilidade de compostos bioativos em polpa congelada de amora-preta (Rubus fruticosus) cv. Tupy [*Stability of bioactive compounds in frozen blackberry (Rubus fruticosus) cv. Tupy*]." *Química Nova* 33 (2010): 1720-1725.

[19] Pillai, Dipin S., P. Prabhasankar, B. S. Jena, and C. Anandharamakrishnan. "Microencapsulation of Garcinia cowa fruit extract and effect of its use on pasta process and quality." *International Journal of Food Properties* 15, no. 3 (2012): 590-604.
[20] Muriach, María, Miguel Flores-Bellver, Francisco J. Romero, and Jorge M. Barcia. "Diabetes and the brain: oxidative stress, inflammation, and autophagy." *Oxidative Medicine and Cellular Longevity* 2014 (2014).
[21] Cepas, Vanesa, Massimo Collino, Juan C. Mayo, and Rosa M. Sainz. "Redox signaling and advanced glycation endproducts (AGEs) in diet-related diseases." *Antioxidants* 9, no. 2 (2020): 142.
[22] Onyango, Arnold N. "Cellular stresses and stress responses in the pathogenesis of insulin resistance." *Oxidative Medicine and Cellular Longevity* 2018 (2018).
[23] Andrade, Priscilla Macedo Lima, Luciana Baptista, Julyana Stoffel Britto, Ana Paula Trovatti Uetenabaro, and Andréa Miura da Costa. "Co-production of tannase and gallic acid by a novel Penicillium rolfsii (CCMB 714)." *Preparative Biochemistry and Biotechnology* 48, no. 8 (2018): 700-706.
[24] Sharma, Kanti Prakash, P. J. John, Pawas Goswami, and Manish Soni. "Enzymatic synthesis of gallic acid from tannic acid with an inducible hydrolase of Enterobacter spp." *Biocatalysis and Biotransformation* 35, no. 3 (2017): 177-184.
[25] Shakeri, Abolfazl, Mohammad R. Zirak, and Amirhossein Sahebkar. "Ellagic acid: a logical lead for drug development?" *Current Pharmaceutical Design* 24, no. 2 (2018): 106-122.
[26] Chhikara, Navnidhi, Ravinder Kaur, Sundeep Jaglan, Paras Sharma, Yogesh Gat, and Anil Panghal. "Bioactive compounds and pharmacological and food applications of Syzygium cumini–a review." *Food & Function* 9, no. 12 (2018): 6096-6115.
[27] Naveed, M., BiBi, J., Kamboh, A. A., Suheryani, I., Kakar, I., Fazlani, S. A., FangFang, X., kalhoro, S. A., Yunjuan, L., Kakar, M. U., Abd El-Hack, M. E., Noreldin, A. E., Zhixiang, S., LiXia, C., & XiaoHui, Z. "Pharmacological values and therapeutic properties of black tea (Camellia sinensis): A comprehensive overview." *Biomedicine & Pharmacotherapy* 100 (2018): 521-531.
[28] Sharma, Kanti Prakash, P. J. John, Pawas Goswami, and Manish Soni. "Enzymatic synthesis of gallic acid from tannic acid with an inducible hydrolase of Enterobacter spp." *Biocatalysis and Biotransformation* 35, no. 3 (2017): 177-184.
[29] Shakeri, Abolfazl, Mohammad R. Zirak, and Amirhossein Sahebkar. "Ellagic acid: a logical lead for drug development?" *Current pharmaceutical design* 24, no. 2 (2018): 106-122.Kawabata K., Yoshioka Y., Terao J. Role of Intestinal Microbiota in the Bioavailability and Physiological Functions of Dietary Polyphenols. *Molecules*. 2019; 24:370.
[30] Abdou, Ebtsam Mohmmed, and Marwa M. Masoud. "Gallic acid–PAMAM and gallic acid–phospholipid conjugates, physicochemical characterization and *in vivo* evaluation." *Pharmaceutical Development and Technology* 23, no. 1 (2018): 55-66.
[31] Bhattacharyya, Sauvik, Sk Milan Ahammed, Bishnu Pada Saha, and Pulok K. Mukherjee. "The gallic acid–phospholipid complex improved the antioxidant

potential of gallic acid by enhancing its bioavailability." *Aaps Pharmscitech* 14, no. 3 (2013): 1025-1033.

[32] D'souza, Jason Jerome, Prema Pancy D'souza, Farhan Fazal, Ashish Kumar, Harshith P. Bhat, and Manjeshwar Shrinath Baliga. "Anti-diabetic effects of the Indian indigenous fruit Emblica officinalis Gaertn: active constituents and modes of action." *Food & Function* 5, no. 4 (2014): 635-644.

[33] Kahkeshani, Niloofar, Fatemeh Farzaei, Maryam Fotouhi, Seyedeh Shaghayegh Alavi, Roodabeh Bahramsoltani, Rozita Naseri, Saeideh Momtaz, Zahra Abbasabadi, Roja Rahimi, Mohammad Hosein Farzaei and Anupam Bishayee. Pharmacological effects of gallic acid in health and disease: A mechanistic review." *Iranian Journal of Basic Medical Sciences* 22, no. 3 (2019): 225-237.

[34] Fernandes, Felipe Hugo Alencar, and Hérida Regina Nunes Salgado. "Gallic acid: review of the methods of determination and quantification." *Critical Reviews in Analytical Chemistry* 46, no. 3 (2016): 257-265.

[35] Brewer, M. S. "Natural antioxidants: sources, compounds, mechanisms of action, and potential applications." *Comprehensive Reviews in Food Science and Food Safety* 10, no. 4 (2011): 221-247.

[36] Saeed, Naima, Muhammad R. Khan, and Maria Shabbir. "Antioxidant activity, total phenolic and total flavonoid contents of whole plant extracts Torilis leptophylla L." *BMC Complementary and Alternative Medicine* 12, no. 1 (2012): 1-12.

[37] Kosuru, Rekha Yamini, Amrita Roy, Sujoy K. Das, and Soumen Bera. "Gallic acid and gallates in human health and disease: do mitochondria hold the key to success?" *Molecular Nutrition & Food Research* 62, no. 1 (2018): 1700699.

[38] Su, Tzu-Rong, Jen-Jie Lin, Chi-Chu Tsai, Tsu-Kei Huang, Zih-Yan Yang, Ming-O. Wu, Yu-Qing Zheng, Ching-Chyuan Su, and Yu-Jen Wu. "Inhibition of melanogenesis by gallic acid: Possible involvement of the PI3K/Akt, MEK/ERK and Wnt/β-catenin signaling pathways in B16F10 cells." *International Journal of Molecular Sciences* 14, no. 10 (2013): 20443-20458.

[39] Sawa, Tomohiro, Mayumi Nakao, Takaaki Akaike, Kanji Ono, and Hiroshi Maeda. "Alkylperoxyl radical-scavenging activity of various flavonoids and other phenolic compounds: implications for the anti-tumor-promoter effect of vegetables." *Journal of Agricultural and Food Chemistry* 47, no. 2 (1999): 397-402.

[40] Shao, Dongyan, Jing Li, Ji Li, Ruihua Tang, Liu Liu, Junling Shi, Qingsheng Huang, and Hui Yang. "Inhibition of gallic acid on the growth and biofilm formation of Escherichia coli and Streptococcus mutans." *Journal of Food Science* 80, no. 6 (2015): M1299-M1305.

[41] Sorrentino, Elena, Mariantonietta Succi, Luca Tipaldi, Gianfranco Pannella, Lucia Maiuro, Marina Sturchio, Raffaele Coppola, and Patrizio Tremonte. "Antimicrobial activity of gallic acid against food-related Pseudomonas strains and its use as biocontrol tool to improve the shelf life of fresh black truffles." *International Journal of Food Microbiology* 266 (2018): 183-189.

[42] Fu, Rao, Yuting Zhang, Tong Peng, Yiran Guo, and Fang Chen. "Phenolic composition and effects on allergic contact dermatitis of phenolic extracts Sapium sebiferum (L.) Roxb. leaves." *Journal of Ethnopharmacology* 162 (2015): 176-180.

[43] Hyun, Ki Hyeob, Ki Cheol Gil, Sung Gun Kim, So-young Park, and Kwang Woo Hwang. "Delphinidin Chloride and Its Hydrolytic Metabolite Gallic Acid Promote Differentiation of Regulatory T cells and Have an Anti-inflammatory Effect on the Allograft Model." *Journal of Food Science* 84, no. 4 (2019): 920-930.

[44] Radan, Maryam, Mahin Dianat, Mohammad Badavi, Seyyed Ali Mard, Vahid Bayati, and Gholamreza Goudarzi. "In vivo and *in vitro* evidence for the involvement of Nrf2-antioxidant response element signaling pathway in the inflammation and oxidative stress induced by particulate matter (PM10): the effective role of gallic acid." *Free Radical Research* 53, no. 2 (2019): 210-225.

[45] Wang, Xinhua, Hongqing Zhao, Chenhui Ma, Lei Lv, Jinping Feng, and Shuguang Han. "Gallic acid attenuates allergic airway inflammation via suppressed interleukin-33 and group 2 innate lymphoid cells in ovalbumin-induced asthma in mice." In *International Forum of Allergy & Rhinology*, vol. 8, no. 11, pp. 1284-1290. 2018.

[46] Dludla, Phiwayinkosi V., Bongani B. Nkambule, Babalwa Jack, Zibusiso Mkandla, Tinashe Mutize, Sonia Silvestri, Patrick Orlando, Luca Tiano, Johan Louw, and Sithandiwe E. Mazibuko-Mbeje. "Inflammation and oxidative stress in an obese state and the protective effects of gallic acid." *Nutrients* 11, no. 1 (2018): 23.

[47] Variya, Bhavesh C., Anita K. Bakrania, and Snehal S. Patel. "Antidiabetic potential of gallic acid from Emblica officinalis: Improved glucose transporters and insulin sensitivity through PPAR-γ and Akt signaling." *Phytomedicine* 73 (2020): 152906.

Chapter 7

Gallic Acid: A Potential Anti-Tumor Agent

Pranali Pangam[*]
Swapnali Patil[†]
Poournima Sankpal[‡]
and Sachinkumar Patil[§]

Ashokrao Mane College of Pharmacy,
Peth-Vadgaon, Kolhapur, Maharashtra, India

Abstract

Tumors are the second biggest global cause of mortality. The latest research in the area is concerned with how cancer therapy resistance arises and how to combat or prevent it. According to the present predicament, innovative anti-cancer drugs are urgently required for treatment of cancer cells that are resistant to chemotherapy. Phytochemical's pharmacological properties and ability to target a variety of biological pathways play a crucial part in the development of cancer therapies.

For the prevention and treatment of tumors, natural phenolic substances such as gallic acid (3,4,5 trihydroxybenzoic acid; GA) gained popularity. Gallic acid (GA) is a polyhydroxy phenolic molecule, typically found in natural sources including amla, berries, grapes, apple peels, green tea, gallnut, sumac, tea leaves, wine, chestnut, and oak bark. The most recent research on gallic acid's anti-tumor properties in various malignancies was examined, with an emphasis on the molecular

---

[*] Corresponding Author's Email: pranalipangam27@gmail.com.
[†] Corresponding Author's Email: swapnali30may99@gmail.com.
[‡] Corresponding Author's Email: poornima6@gmail.com.
[§] Corresponding Author's Email: sachinpatil.krd@gmail.com.

In: The Chemistry of Gallic Acid and Its Role in Health and Disease
Editor: Jeff C. Murdoch
ISBN: 979-8-88697-672-4
© 2023 Nova Science Publishers, Inc.

mechanisms and cellular pathways that lead to tumor cell apoptosis and migration. When gallic acid and chemotherapeutic drugs are administered simultaneously, tumor proliferation is suppressed more effectively.

Keywords: gallic acid, tumor, polyhydroxy phenolic substance

Introduction

In 2022, 1,918,030 new cancer cases and 609,360 cancer deaths are projected to occur in the United States, including approximately 350 deaths per day from lung cancer, the leading cause of cancer death. Approximately 105,840 of the 130,180 lung cancer deaths (81%) in 2022 will be caused by cigarette smoking directly, with an additional 3650 due to second-hand smoke. Breast (2.26 million instances), colon and rectum (1.93 million instances), prostate (1.41 million instances), skin (non-melanoma) (1.20 million instances), and stomach (1.03 million instances) were the most frequent tumor types in 2020 (1.09 million Instances). Lung (1.80 million fatalities), colorectal (916,000 deaths), hepatic (830,000 deaths), gastric (769,000 deaths), and breast (685,000 deaths) were the most common tumor cases cause death. A third of tumor-related fatalities are attributed to smoking, having a high body mass index, and consuming alcohol. About 30% of tumor instances in low- and lower-middle-income nations are caused by illnesses including hepatitis and the human papillomavirus (HPV) (Ali, Anwar, et al. 2022). Many tumors can be cured if discovered early and treated efficiently. Surgical removals of the tumor, radiation therapy, and/or systemic therapies (chemotherapy, hormonal treatments, and targeted biological therapies) are the most common forms of treatment. Lack of targeted medication to the tumor tissue is one of the limitations of current tumor therapy (Shahid Rizwana, Sadia Khan, et al. 2022).

Natural phenolic compounds have gained prominence in the prevention and treatment of tumors, but it is still unclear how exactly they work. The polyhydroxy phenolic acid such as gallic acid (GA) can only be obtained naturally from amla, berries, grapes, apples, green tea, gallnut, sumac, tea leaves, wine, chestnut, and oak bark. Phytochemicals are crucial in the development of cancer treatments due to their pharmacological activities and ability to target different molecular pathways. They also play a significant role in the treatment of tumors (Shahid Rizwana, Sadia Khan, et al. 2022). The expression of molecular pathways like PI3K/Akt is important in the progression of cancer that is significantly suppressed by gallic acid. Natural compounds made from plants have a low bioavailability, which reduces their capacity to fight tumors. However, administering these substances in nanotechnology-based formulations or even combining them with other phytochemicals can enhance their bioavailability. Additionally, the idea of combining polyphenols with traditional chemotherapy medicines offers significant benefits. When gallic acid is administered along with chemotherapy medications, cancer malignancy is suppressed more effectively. For the targeted delivery of gallic acid at the tumor site, various nano-vehicles including organic and inorganic nanomaterials have also been created (Samad N. and A. Javed, 2018).

Gallic Acid

Gallic acid (GA), a naturally occurring low molecular weight triphenolic chemical, is a common free compound or component of tannins in the plant kingdom (Patil P. and S. Killedar, et al., 2021). Gallnuts, sumac, witch hazel, watercress, tea leaves, areca nuts, bearberry (Arctostaphylos), blackberry, and Caesalpinia Mimosoideae are some plants that contain gallic acid. They can be found as methylated, galloyl, polygalloyl, glucose esters, or glycerol derivatives of catechin. Gallic acid is easily produced by hydrolyzing the large quantities of tannin that plants produce in alkaline or acidic solutions. Alkyl gallates are frequently added to foods as antioxidant supplements (Patil, P. and S. Killedar, et al., 2022). Gallic acid has reportedly attracted the interest of different chemical and pharmaceutical sectors due to its many intriguing features and practical uses (Gao Tiyu, Yunxiang Ci, Hongyuan Jian, et al. 2000).

In numerous illnesses, such as cancer, neurological disorders, and ageing, gallic acid exhibits a variety of biological functions. Accordingly, gallic acid could be considered as a promising lead compound for development of new drug. Gallic acid has been studied for its antioxidant, anticancer, antibacterial, antifungal, antiviral, anti-inflammatory, and anti-diabetic properties. In fact, compounds with a greater number of hydroxylic groups consist of better anticancer activity as compared to those with a lower number. In this regard, gallic acid possesses three hydroxyl groups attached to three, four, and five positions of a benzoic acid core, which has been reported to be more effective than other phenols (Zahrani, Nourah A., et al. 2020).

They function as antioxidants and can prevent the oxidation of proteins, DNA, lipids, and enzymes that produce free radicals. The natural antioxidant defence system normally balances the production of reactive oxygen species, and oxidative stress is produced when this balance favours the ROS. The pharmacokinetic properties of gallic acid, such as its large particle size, restricted absorption, poor solubility, low bioavailability, and quick elimination, negatively affect how well it is taken and utilised by people (Sanchez Martin, Victoria, et al. 2022).

Red wines and green teas are processed drinks that contain gallic acid, a naturally occurring polyphenolic compound (3, 4, 5-trihydroxybenzoic acid). It emerges in plants as free acids, esters, catechin derivatives, free tannins, and hydrolyzable tannins. Tannic acid can be hydrolyzed using an acid, an alkali, or a microbial tannase to create gallic acid. Oxidation makes it simple to separate gallic acid from gallotannins. The fastest way to do it is to use concentrated sulfuric acid to precipitate it out of an aqueous solution. Allowing water to passively oxidise with ambient oxygen is a slower way to create gallic acid. In the pharmaceutical sector, it is mostly utilised for the manufacture of antibacterial medications like trimethoprim. The chemical synthesis of food preservatives like pyrogallol and gallates in the food industry uses gallic acid as a substrate (Mukherjee, Gargi, et al. 2003).

Gallates are often recognised as the ester derivatives of gallic acid in a variety of plants, and their biological properties are also being studied. Gallic acid ester derivatives have been found to have anticancer properties. According to these investigations, gallates caused apoptosis in several cancer cell lines. According to structure activity relationship (SAR) studies, gallic acid derivatives can serve as an antioxidant since they contain hydroxyl groups (Locatelli, Claudriana, Fabiola, et al. 2013). It was discovered that the para-substituted-OH group had a powerful radical scavenger. Antioxidant activity is influenced by intramolecular hydrogen bonding as well as by hydroxyl groups. For instance, the readily ionizable carboxylic group in phenolic acids is responsible for their efficient hydrogen donation ability (Subramanian A. P., A. A. John, et al. 2015).

The primary factor contributing to the effectiveness of gallic acid as an antioxidant as compared to pyrogallo-5 is that it possesses a readily ionizable carboxyl group, which would result in an effective hydrogen donation propensity of phenolic acids (Sroka Z. and W. Cisowski., 2003). Therefore, gallic acid is a key component in developing new effective pharmacophores due to its bioactivity and abundant availability in nature. Numerous physiologically and pharmacologically active molecules have recently been produced as a result of improvements in the preparation of gallic acid derivatives. Additionally, research has shown that gallic acid and its derivatives have the ability to specifically trigger the apoptosis of cancer cells without affecting healthy cells. Due to chemical modifications made to the gallic acid molecule, several physicochemical properties exist, such as the lipid solubility in the alkyl esters, 3,4,5-triacylated benzoic acid, and its esters (Badhani, Bharti, Neha Sharma, 2015).

It has been noted that the efficiency of antioxidant substances can function well in both aqueous and lipid system. Examples in this regard include the fact that dodecyl gallate 7c (a food additive) inhibits the enzyme xanthine oxidase more effectively than gallic acid. The amphipathic property of these ester derivatives has been attributed to these biological activities (Hejchman, Elżbieta, Przemysław Taciak, et al. 2015). Gallic acid has undergone a number of structural changes to access its structure-activity relationships. This study seeks to provide an update on the chemistry surrounding pharmacophores containing gallic acid as well as their therapeutic effects (Subramanian A. P., A. A. John, et al. 2015).

Gallic Acid as an Anti-Tumor Agent

Ovarian Carcinoma

Ovarian cancer is the most familiar gynaecologic cancer among ladies and the ninth most widespread malignancy in the US. In developed regions, it is one of the major contributing factors of tumor-related mortality in females (Kubo, Isao, Ken-ichi Fujita, et al. 2003). Gallic acid has demonstrated the best inhibitory treatment on human ovarian malignancy cells among eight characteristic phenolic compounds (Forester, Sarah C., Ying Y, et al. 2014). Gallic acid has an inhibitory impact on two ovarian disease cell lines such as A2780/CP70 and OVCAR-3 than ordinary ovarian cell line such as IOSE-364. It reportedly inhibits vascular endothelial growth factor (VEGF) release and especially decreases the proliferation of tumor cells. This polyphenol increases PTEN (Phosphate and TENsin homolog) expression by decreasing AKT phosphorylation and hypoxia inducible factor 1 (HIF-1) protein expression (Engelender, Simone, et al. 2017). Gallic acid's inhibitory effect on in-vitro angiogenesis and VEGF expression is represented by the PTEN /AKT /HIF-1 pathway. Gallic acid and its derivatives found in amla extract prevent apoptotic cell death while significantly increasing the expression of the autophagic proteins such as LC3B-II and beclin1.It is reported that, the gallic acid reduces the expression of several angiogenic genes, such as hypoxia inducible factor 1α (HIF-1α) both in OVCAR3 and SW626 cells (Ferlay, Jacques, Isabelle Soerjomataram, et al. 2012).

A prior investigation on the three ovarian cell lines such as OVCAR-8, A2780, and A2780cis demonstrates that the gallic acid shows anti-ovarian cancer effects. These findings provide strong evidence for the great efficacy of gallic acid in the detection and management of ovarian cancer (Forester, Sarah C., Ying Y, et al. 2014).

Colon/Colorectal Cancer

In many countries, particularly in the western regions, colon-rectal cancer is one of the most common and well-known health problems. According to studies by the World Health Organization, around one million people are diagnosed with colon cancer each and every year (Ghanemi, Fatima Zahra, et al. 2017). Gallic acid has a significant impact on colon cancer. As the concentration of gallic acid rises, human colon tumor HCT-15 cells undergo necrosis or apoptosis, which slows cell proliferation (Gil-Longo, Jose, et al. 2010).

In addition, gallic acid and 3-O-methylgallic acid have the ability to reduce colorectal cancer (CRC) and viability of Caco-2 cell. This decline is mostly caused by gallic acid and 3-O-methylgallic acid's capacity to inhibit the cell cycle at the G0/G1phase. Gallic acid and 3-O-methylgallic acid's ability to inactivate the translational regulators NF-kB, AP-1, OCT-1, and STAT-1 partially mediates their anti-tumor effects. It produces a dose dependent cytotoxic effect through the introduction of apoptosis on CRC cell lines. CLP initiate PARP cleavage in CT-26 and HCT-116 cells and natural apoptotic pathway through the caspase-9 application. Additionally, CLP promote p53 activation, which causes cell cycle arrest at the G1 stage (Dragicevic, Natasa, et al. 2011).

Lymphoma (Lymphatic Cancer)

Gallic acid caused lymphoma cell line apoptosis in a dose-dependent manner. After gallic acid therapy, the proliferating cell nuclear antigen and IkappaB kinase (I-B) protein were downregulated and the NF-B protein was upregulated. These findings suggest that gallic acid may be used as a chemotherapeutic treatment for lymphoma. One of the well-known biochemical alterations during apoptosis is the DNA fragmentation of gallic acid and its ester derivatives. Additionally, there were other morphological modifications such as cell shrinkage, chromatin condensation, and the presence of apoptotic bodies that resulted in cell death (Subramanian A. P., A. A. John, et al. 2015).

Prostate Cancer

The main anticancer substance, gallic acid, inhibited the development of DU145 prostate cancer cells. ROS production and mitochondria-mediated apoptosis have a role in the decrease in the viability of DU145 cells. Through the activation of Chk1 and Chk2 and the inhibition of Cdc25C and Cdc2, gallic acid caused the cell cycle to stop at the G2/M stages. The autoxidation of gallic acid successfully eliminated the cancerous prostate cells. The amount of ROS also increased noticeably as a result of this autoxidation. Caspases 3, 8, and 9 were activated as a result of the release of cytochrome c and loss of mitochondrial potential. Prostate cancer cells demonstrated a dose-dependent apoptosis in response to gallic acid. The impact of gallic acid on PC3 prostate cancer cells not only produced DNA damage but also prevented DNA repair by changing the DNA repair genes (Hsu, Jeng-Dong, Shao-Hsuan Kao, et al. 2011).

Breast Cancer

Human breast cancer cell MCF-7 cell proliferation is significantly and dose-dependently inhibited by gallic acid therapy. Gallic acid significantly delayed G2/M phase but only moderately affect the population of sub-G1MCF-7 cells. The S-G2/M transitional protein levels of cyclins, cyclin-dependent kinases (CDKs), and their regulatory proteins were studied. (Chen, Huei-Mei, Yang-Chang Wu, et al. 2009) According to this study, gallic acid treatment boosted the levels of the negative regulators such as p27Kip1 and p21Cip1, while it decreased the levels of cyclin A, CDK2, cyclin B1, and cdc2/CDK1. Additionally, gallic acid treatment decreases Skp2, a particular ubiquitin E3 ligase for polyubiquitination of p27Kip1. Further study revealed that gallic acid reduced the interaction of Skp2-p27Kip1 and the polyubiquitination of p27Kip1 in MCF-7 cells. Additionally, the accumulation of G2/M phase caused by gallic acid was significantly reduced by p27Kip1 knockdown (Kang Nalae, Ji-Hyeok Lee, et al. 2015). These results suggest that gallic acid may increase the level of p27Kip1 through the breakdown of the p27Kip1/Skp2 connection and the subsequent degradation of p27Kip1 by the proteosome, resulting in the arrest of the MCF-7 cell in the G2/M phase. It is suggested that gallic acid should be beneficial for the management of breast cancer and p27Kip1-deficient carcinomas (Latha R. Cecily Rosemary, et al. 2011).

Lung Cancer

Lung cancer is the most common reason for malignancy passing around the world. According to the present predicament, the most well-known cause of cancer deaths worldwide is lung cancer, As per united state study reports, it is estimated that there will be 1,918,030 new cancer cases and 609,360 cancer deaths in 2022, with lung cancer being the primary cause of death accounting for around 350 of those fatalities daily. Clinical interventions have had very little chance of reducing lung cancer-related deaths in recent years. More than 80% of occurrences of lung cancer are non-small-cell lung cancer (NSCLC), which is the most prominent subtype of lung cancer. Gallic acid triggers apoptosis in EGFR (epidermal growth factor receptor) mutant non-small cell lung cancer (NSCLC) cells. According to studies, this polyphenol treatment lowers EGFR levels, which are crucial for NSCLC survival (Lee Jin-Ching, Wei-Chun Chen, et al. 2011).

It initiates the process of EGFR turnover, which causes apoptosis in EGFR-TKI (tyrosine kinase inhibitor)-resistant cell lines that are susceptible to EGFR motioning for survival. In short, gallic acid can accelerate EGFR turnover, which triggers apoptosis in EGFR-dependent lung cancers which depends on EGFR for growth and survival. Furthermore, it shows antitumorigenic effects in non-small cell lung cancer (NSCLC) that is resistant to Tyrosine Kinase Inhibitor (TKI) therapy. Gallic acid treatment inhibits Src-Stat3-mediated signaling and decreases the expression of Stat3-regulated tumor. advancing genes, eventually triggering apoptosis and cell cycle arrest in the TKI-resistant lung cancer. This information reveals a role for Src-Stat3 flagging in gallic acid mediated tumor suppression movement and, more importantly, provides a unique understanding of gallic acid for reducing TKI-resistant lung growth (Lindenbach Brett D. and Charles M. Rice. 2013).

Oxidative mechanisms and caspase activation were both seen to be involved in the apoptotic process. These results further raise the prospect of using gallic acid in lung cancer treatment, particularly to overcome medication resistance. In connection to reactive oxygen species (ROS) and glutathione, gallic acid demonstrated anti-tumor effects on Calu-6 and A549 lung cancer cells (GSH). In lung fibroblast cells, gallic acid was reported to cause c-Jun nh2-terminal kinase-dependent apoptosis that was triggered by reactive oxygen species. Gallic acid-mediated hydrogen peroxide generation was discovered to be the JNK signaling pathways starter, which was followed by the activation of the p53 pathway, which causes apoptosis. Gallic acid induced morphological modifications such as cell rounding and cell shrinkage. Additionally, the treatment with gallic acid reduced the potential of the mitochondrial membrane and raised intracellular reactive oxygen species, which triggered caspase-3. Contrarily, the absence of caspase-8 activation suggests that the intrinsic mechanism of cell death is involved. Gallic acid caused apoptosis in a dose and time-dependent manner (You Bo Ra, and Woo Hyun Park. 2010).

Oral Cancer

The plant polyphenol gallic acid has been reported to have cytotoxic and pro-apoptotic effects on human oral cancer HSC-2 cells. Gallic acid sensitivity was higher in human oral cancer HSC-2 cells compared to healthy human gingival fibroblasts. Gallic acid raised the amount of intracellular reactive oxygen species, induced lipid peroxidation, and reduced intercellular glutathione. The HSC-2 cells underwent concentration-dependent apoptosis. Overall, the cytotoxicity was brought on by the treatment's elevation of oxidative stress, which resulted in cell death. TNF-, TP53BP2, and GADD45A were upregulated along with the anti-apoptotic genes Survivin and cIAP1, which were downregulated. This demonstrated that gallic acid causes HOSCC cells to undergo apoptosis. By inhibiting the translocation of NF-B and RhoA from the cytosol to the nucleus, gallic acid decreased the migration and invasion of SCC4 cells. Matrix metalloproteinase (MMP)-2 and MMP-9 activities were also inhibited. This supported the use of gallic acid as a treatment for oral cancer (Chia Yi-Chen, Ranjan Rajbanshi, et al. 2010).

Esophageal Cancer

In TE-2 cells, gallic acid showed a substantial antiproliferation effect. An increase in the pro-apoptosis Bax protein activity in cancer cells was the biochemical mechanism responsible for gallic acid induced apoptosis. On the other hand, gallic acid downregulated the survival Akt/mTOR pathway as well as anti-apoptosis proteins including Bcl-2 and Xiap. In contrast, non-cancerous cells expressed pro-apoptosis-related proteins later. After 12 hours of treatment, the TE-2 cells showed notable morphological abnormalities that weren't seen in the normal CHEK-1 cells (Gao Ying, Wei Li, Lingyan Jia, 2013).

Cervical Cancer

Human umbilical vein endothelial cells (HUVEC) and HeLa cervical cancer cells were used to examine the effects of gallic acid on cell death and growth inhibition. In both types of cells, gallic acid was observed to cause cell lysis. The reduction of mitochondrial membrane potential in the cervical cancer cells occurred concurrently with this cell demise. Gallic acid treatment of HeLa cells resulted in an increase in ROS production and GSH depletion. Gallic acid significantly reduced GSH levels, lost mitochondrial membrane potential, and increased ROS levels, notably $O(2)(\bullet-)$, all of which impeded the development of HeLa cells in a dose-dependent manner. According to reports, gallic acid inhibits the growth, invasion, and angiogenesis of human cervical cancer cells. A dose-dependent reduction in cell viability was seen after gallic acid treatment of human cancer HeLa cells. Additionally, gallic acid was found to reduce HeLa cell growth. Gallic acid didn't have as much of a cytotoxic effect on healthy HUVECs as it did on cervical cancer cells HeLa and HTB-35 (Zhao B, Hu M. et al. 2013).

Leukemia

After gallic acid therapy, it was clear that leukaemia cells undergo apoptosis, which was followed by a G0/G1 cell cycle. Gallic acid resulted in the suppression of ribonucleotide reductase, DNA damage, and fragmentation in cancer cells. The apoptosis is linked to the mitochondrial pathway by encouraging the release of cytochrome c, apoptosis-inducing factor (AIF), and endonuclease G (Endo G), as well as by upregulating the Bcl-2 protein, which activates caspases 4, 9, and 3 (Locatelli, Claudriana, Rober Rosso, et al. 2008). It was discovered that ROS formation contributed to the apoptosis of gallic acid induced promyelocytic leukaemia HL-60RG cell death. In the gallic acid treated HL 60 cells, the production of ROS was dose dependent. After gallic acid treatment, the ability to trigger apoptosis was closely associated with the intracellular peroxide level. The activity of intracellular peroxide level was less active than the role of ROS formation in triggering apoptosis. In addition, NF-B and BCR/ABL tyrosine kinase were suppressed by gallic acid. As BCR/ABL kinase, NF-B activity, and COX-2 are inhibited, cell death brought on by gallic acid therapy includes death receptor and mitochondrial-mediated pathways. On murine lymphoblastic L1210 leukaemia cells, the ester derivatives of gallic acid were able to cause apoptosis by the DNA ladder fragmentation pattern. Additionally, GSH was depleted in the mitochondria and cytoplasm, and NF-kappaB was activated (Isuzugawa Kazuto, Makoto Inoue, et al. 2001).

Hepatocellular Carcinoma (HCC)

Nearly 10.9 million new instances of cancer and 67 million cancer-related deaths are reported each year, which indicates that the incidence rate of cancer is significantly rising. The most prevalent illness among men globally is considered to be liver cancer. Recent surgical resections and promising diagnostic techniques have significantly improved the prognosis for HCC. The significant risk of recurrence even after hepatic resection (50%–70% after 5 years) prevents the long-term survival rate from being adequate. Furthermore, there are currently no proven effective systemic chemotherapeutic treatments for HCC. New treatment approaches are hence need for HCC retraction (Wang, Yingqiang, Qianqian Luo, et al. 2014).

Natural products have received a lot of attention in the last 30 years because of their intriguing capabilities to treat cancer. They therefore become excellent prospects for creating fresh chemopreventive and anticancer drugs. The current setup suggested that gallic acid treatment might modify the activity of tumor markers and prevent HCC (Singh, Brahma N., Braj R. Singh, et al. 2009).

The administration of gallic acid to the HCC group considerably reduced the rise in blood AFP levels. Gallic acid therapy, which is known to affect the transcription of AFP, is thought to reduce COX-2 gene expression, which is the hypothesised method by which gallic acid might restore AFP blood level. It has been demonstrated that gallic acid decreases COX-2 gene expression via reducing NF-B activity in colon tissue. Moreover, gallic acid therapy would result in a drop in the protein synthesis by hepatic tumor cells due to its anticancer activity, which may therefore be responsible for tumor growth shrinkage and such regression in AFP production by tumor cells (Rosmorduc, Olivier, et al. 2011).

Gallic acid inhibits nuclear translocation of p65-NF-B as well as degradation of inhibitory protein IB and limits the activation and nuclear accumulation of phosphorylated STAT3. The study also noted that gallic acid greatly reduces the level of IL-6 gene expression, which prevents STAT3 from being activated. According to these findings, gallic acid reduces inflammation through inhibiting NF-B and IL-6/p-STAT3 activation (Ji Yuan-Yuan, Zhi-Dong Wang et al. 2013).

Melanoma Cells

In A375S2 human melanoma cells, the natural antioxidant gallic acid significantly reduces cell growth and induces apoptosis. Gallic acid treatment resulted in a dose and time-dependent drop in the proportion of viable cells (Lo Chyi, Tung-Yuan Lai, et al. 2013). The proapoptotic Bax proteins were upregulated in the molecular process of apoptosis, whereas the antiapoptotic Bcl-2 proteins were downregulated. Gallic acid caused cytosolic release of cytochrome c, which promoted the activation of caspase-9 and caspase-3 and eventually resulted in apoptotic cell death. It also lowered the amount of mitochondrial membrane potential in a time-dependent way (Subramanian V., Venkatesan B., et al. 2011). Gallic acid also encouraged the release of endonuclease G and the apoptosis-inducing factor (AIF) (Endo G). As a result, a caspase-independent mechanism was used to trigger apoptosis. This provided evidence for the antimetastatic properties of gallic acid. Additionally, this was connected to the Ras, p-ERK signalling pathways, which resulted in the suppression of MMP-2 in A375S2 human melanoma cells (Locatelli, Claudriana, Paulo C et al. 2009). On B16F10 cells, the octyl, decyl, dodecyl, and tetradecyl gallates caused apoptosis, which resulted in cell death. Free radicals were produced, glutathione (GSH) and ATP were both depleted, NF-kappaB was activated, and cell adhesion was inhibited as a result of the gallate treatment. GSH depletion was significantly aided by the activity of gamma-glutamyl cysteine synthase. The result of oxidative stress, which had various mechanisms, was the suppression of growth. By inhibiting the expression of the adhesion proteins ICAM-1 and VCAM-1, the octyl, dodecyl, and tetradecyl gallates also have important effects on cell adhesion and migration (Ortega E., M. C. Sadaba, A. I. et al. 2003).

Bladder Cancer

One of the most prevalent malignant tumors of the urinary system, bladder cancer has a terrible prognosis and a low survival rate. Patients with bladder cancer may be in the middle or late stages of the illness because they have clinical signs. It is really simple to relapse in this situation, even after obtaining treatment. In order to significantly increase bladder cancer patients' chances of survival, it is imperative to find a safe and efficient medicine. More researchers are now focused on developing novel, powerful anticancer medications from natural sources (Schulz G. B., Stief C. G., et al. 2019). Tannins can cause the apoptosis of bladder cancer T24 cells. The T24 cells viability is most strongly inhibited by gallic acid. Gallic acid also has an impact on the shape of T24 cells, greatly reduces cell proliferation, interrupts the S phase of the cell cycle, and induces cell death, primarily in the early stages. Therefore, the emphasis of the ensuing study was on the molecular mechanism through which gallic acid promotes the death of T24 cells. Apoptosis, which is connected to ROS build up and MMP depolarization, is promoted by dysfunctional mitochondria, according to a number of studies (Sivaraman Siveen, Kodappully, Kirti Prabhu, et al. 2017). Additionally, ROS production and MMP depolarization have been shown to alter the intracellular milieu, result in oxidative damage, and even trigger cell death (Song Bingbing, Jia Li, et al. 2016).

After treatment with gallic acid, Bax, P53, Caspase-3, and Cyt-c proteins and gene expression levels considerably rose, but Bcl-2 significantly decreased. The mitochondrial dysfunction that gallic acid caused in T24 cells may be related to that condition. Gallic acid was also discovered to prevent the spread of gastric cancer AGS cells by inhibiting NF-kB and down-regulating PI3K/Akt/small GTPase signaling. The transduction of the PI3K/Akt/NF-kB signaling pathway reduces tumor cell death while promoting tumor cell growth and metastasis. As a result, blocking this route could be an effective strategy for treating different tumors. Modern research has confirmed that several Chinese herbal extracts can inhibit PI3K/Akt/NF-kB to increase the death of different tumor cells (Tsai Chung-Lin, Ying-Ming Chiu, et al.

2018). In this signaling pathway, Akt is initially encouraged to become phosphorylated by PI3K, which then either activates or inhibits its downstream substrates, such as Bad, Caspase-9, and NF-kB. P-Akt was thought to help cells survive by controlling the activity of IKKa, an NF-kB inhibitor, and by encouraging the phosphorylation of IkBa, an NF-kB inhibitor (Wang Ruixuan, Lijie Ma, Dan Weng, et al. 2016). In the end, the P-IkBa can release protease and the NF-kB dimer, allowing NF-kB to enter the cell nucleus and start the transcription of the appropriate genes.

These findings suggested that the suppression of the PI3K/Akt/NF-kB signaling pathway may be strongly related to the apoptosis that gallic acid produced in T24 cells. Malignant tumors also have a high mortality rate; around 70–80% of cancer patients pass away as a result of the metastasis and dissemination of tumor cells (Song Bingbing, Jia Li, et al. 2016). Bladder cancer is a frequent malignant tumor of the urinary system, it has high invasive potential, and it may return and spread after surgery. Previous research indicated that gallic acid has anti-metastasis activity *in vitro* and vivo. As a result, we chose the human bladder T24 cell line due to its strong infiltration and metastatic activities (Wen Haiyan, Siqi Zhou, et al. 2019). The migration and invasion of tumor cells are strongly related to their lethality, and the development and metastasis of tumor cells are based on the production of new blood vessels.

More significantly, VEGF is a key mediator in boosting tumor angiogenesis; it may encourage the formation of vascular endothelial cells and trigger vascular proliferation, both of which are directly linked to the development of cancer. According to the relevant research, VEGF can enhance angiogenesis and the spread of cancer cells and is directly associated to metastatic activity (Song Bingbing, Jia Li, et al. 2016). VEGF inhibition is currently a key strategy for the treatment of tumors and a key component of targeted therapy. Additionally, gallic acid dramatically reduced the production of the VEGF protein and prevented the migration and invasion of T24 cells (Roy, Rituparna, Debolina Pal, et al. 2019).

Conclusion

A broad range of illnesses that can affect any region of the body are collectively referred to as cancer. It is among the world's major causes of death. Surgery, chemotherapy, or radiation can cure a major number of malignancies. Chemotherapy is a type of tumor treatment that makes use of chemicals, particularly anticancer medications. It is used both before and after surgery, as well as in conjunction with radiation. This method has a variety of negative side effects and, when cancer is found in advanced stages, it cannot be cured, which encourages ongoing research into anticancer medications. The relevance of natural substances having anticancer properties is increasing. Gallic acid and its derivatives are suggested as one of the top choices for treating cancer in this review.

References

Ali, A., Manzoor, M. F., Ahmad, N., Aadil, R. M., Qin, H., Siddique, R., Riaz, S., Ahmad, A., Korma, S. A., Khalid, W., & Aizhong, L. "The Burden of Cancer, Government Strategic Policies, and Challenges in Pakistan: A Comprehensive Review." *Frontiers in Nutrition* (2022): 1553.

Badhani, Bharti, Neha Sharma, and Rita Kakkar. "Gallic acid: a versatile antioxidant with promising therapeutic and industrial applications." *Rsc Advances* 5, no. 35 (2015): 27540-27557.

Chen, Huei-Mei, Yang-Chang Wu, Yi-Chen Chia, Fang-Rong Chang, Hseng-Kuang Hsu, Ya-Ching Hsieh, Chih-Chen Chen, and Shyng-Shiou Yuan. "Gallic acid, a major component of Toona sinensis leaf extracts, contains a ROS-mediated anti-cancer activity in human prostate cancer cells." *Cancer Letters* 286, no. 2 (2009): 161-171.

Chia, Yi-Chen, Ranjan Rajbanshi, Colonya Calhoun, and Robert H. Chiu. "Anti-neoplastic effects of gallic acid, a major component of Toona sinensis leaf extract, on oral squamous carcinoma cells." *Molecules* 15, no. 11 (2010): 8377-8389.

Dragicevic, Natasa, Adam Smith, Xiaoyang Lin, Fang Yuan, Neil Copes, Vedad Delic, Jun Tan, Chuanhai Cao, R. Douglas Shytle, and Patrick C. Bradshaw. "Green tea epigallocatechin-3-gallate (EGCG) and other flavonoids reduce Alzheimer's amyloid-induced mitochondrial dysfunction." *Journal of Alzheimer's Disease* 26, no. 3 (2011): 507-521.

Engelender, Simone, and Ole Isacson. "The threshold theory for Parkinson's disease." *Trends in Neurosciences* 40, no. 1 (2017): 4-14.

Ferlay, Jacques, Isabelle Soerjomataram, Rajesh Dikshit, Sultan Eser, Colin Mathers, Marise Rebelo, Donald Maxwell Parkin, David Forman, and Freddie Bray. "Cancer incidence and mortality worldwide: sources, methods and major patterns in GLOBOCAN 2012." *International Journal of Cancer* 136, no. 5 (2015): E359-E386.

Forester, Sarah C., Ying Y. Choy, Andrew L. Waterhouse, and Patricia I. Oteiza. "The anthocyanin metabolites gallic acid, 3-O-methylgallic acid, and 2, 4, 6-trihydroxybenzaldehyde decrease human colon cancer cell viability by regulating pro-oncogenic signals." *Molecular Carcinogenesis* 53, no. 6 (2014): 432-439.

Gao, Tiyu, Yunxiang Ci, Hongyuan Jian, and Chengcai An. "FTIR investigation of the interaction of tumor cells treated with caffeic acid and chlorogenic acid." *Vibrational Spectroscopy* 24, no. 2 (2000): 225-231.

Gao, Ying, Wei Li, Lingyan Jia, Bo Li, Yi Charlie Chen, and Youying Tu. "Enhancement of (−)-epigallocatechin-3-gallate and theaflavin-3-3′-digallate induced apoptosis by ascorbic acid in human lung adenocarcinoma SPC-A-1 cells and esophageal carcinoma Eca-109 cells via MAPK pathways." *Biochemical and Biophysical Research Communications* 438, no. 2 (2013): 370-374.

Ghanemi, F. Z., Belarbi, M., Fluckiger, A., Nani, A., Dumont, A., De Rosny, C., Aboura, I., Khan, A. S., Murtaza, B., Benammar, C., Lahfa, B. F., Patoli, D., Delmas, D., Rébé, C., Apétoh, L., Khan, N. A., Ghringhelli, F., Rialland, M., & Hichami, A. "Carob leaf polyphenols trigger intrinsic apoptotic pathway and induce cell cycle arrest in colon cancer cells." *Journal of Functional Foods* 33 (2017): 112-121.

Gil-Longo, José, and Cristina González-Vázquez. "Vascular pro-oxidant effects secondary to the autoxidation of gallic acid in rat aorta." *The Journal of Nutritional Biochemistry* 21, no. 4 (2010): 304-309.

Hejchman, Elżbieta, Przemysław Taciak, Sebastian Kowalski, Dorota Maciejewska, Agnieszka Czajkowska, Julia Borowska, Dariusz Śladowski, and Izabela Mlynarczuk-Bialy. "Synthesis and anticancer activity of 7-hydroxycoumarinyl gallates." *Pharmacological Reports* 67, no. 2 (2015): 236-244.

Hsu, Jeng-Dong, Shao-Hsuan Kao, Ting-Tsz Ou, Yu-Jen Chen, Yi-Ju Li, and Chau-Jong Wang. "Gallic acid induces G2/M phase arrest of breast cancer cell MCF-7 through stabilization of p27Kip1 attributed to disruption of p27Kip1/Skp2 complex." *Journal of Agricultural and Food Chemistry* 59, no. 5 (2011): 1996-2003.

Isuzugawa, Kazuto, Makoto Inoue, and Yukio Ogihara. "Ca2+-dependent caspase activation by gallic acid derivatives." *Biological and Pharmaceutical Bulletin* 24, no. 7 (2001): 844-847.

Ji, Yuan-Yuan, Zhi-Dong Wang, Zong-Fang Li, and Ke Li. "Interference of suppressor of cytokine signaling 3 promotes epithelial-mesenchymal transition in MHCC97H cells." *World Journal of Gastroenterology: WJG* 19, no. 6 (2013): 866.

Kang, Nalae, Ji-Hyeok Lee, WonWoo Lee, Ju-Young Ko, Eun-A. Kim, Jin-Soo Kim, Min-Soo Heu, Gwang Hoon Kim, and You-Jin Jeon. "Gallic acid isolated from Spirogyra sp. improves cardiovascular disease through a vasorelaxant and antihypertensive effect." *Environmental Toxicology and Pharmacology* 39, no. 2 (2015): 764-772.

Kubo, Isao, Ken-ichi Fujita, Ken-ichi Nihei, and Noriyoshi Masuoka. "Non-antibiotic antibacterial activity of dodecyl gallate." *Bioorganic & Medicinal Chemistry* 11, no. 4 (2003): 573-580.

Latha, R. Cecily Rosemary, and P. Daisy. "Insulin-secretagogue, antihyperlipidemic and other protective effects of gallic acid isolated from Terminalia bellerica Roxb. in streptozotocin-induced diabetic rats." *Chemico-Biological Interactions* 189, no. 1-2 (2011): 112-118.

Lee, Jin-Ching, Wei-Chun Chen, Shou-Fang Wu, Chin-kai Tseng, Ching-Yi Chiou, Fang-Rong Chang, Shih-hsien Hsu, and Yang-Chang Wu. "Anti-hepatitis C virus activity of Acacia confusa extract via suppressing cyclooxygenase-2." *Antiviral Research* 89, no. 1 (2011): 35-42.

Lindenbach, Brett D., and Charles M. Rice. "The ins and outs of hepatitis C virus entry and assembly." *Nature Reviews Microbiology* 11, no. 10 (2013): 688-700.

Lo, Chyi, Tung-Yuan Lai, Jai-Sing Yang, Jen-Hung Yang, Yi-Shih Ma, Shu-Wen Weng, Hui-Yi Lin, Hung-Yi Chen, Jaung-Geng Lin, and Jing-Gung Chung. "Gallic acid inhibits the migration and invasion of A375. S2 human melanoma cells through the inhibition of matrix metalloproteinase-2 and Ras." *Melanoma Research* 21, no. 4 (2011): 267-273.

Locatelli, Claudriana, Fabiola Branco Filippin-Monteiro, and Tânia Beatriz Creczynski-Pasa. "Alkyl esters of gallic acid as anticancer agents: a review." *European Journal of Medicinal Chemistry* 60 (2013): 233-239.

Locatelli, Claudriana, Paulo C. Leal, Rosendo A. Yunes, Ricardo J. Nunes, and Tânia B. Creczynski-Pasa. "Gallic acid ester derivatives induce apoptosis and cell adhesion inhibition in melanoma cells: the relationship between free radical generation, glutathione depletion and cell death." *Chemico-Biological Interactions* 181, no. 2 (2009): 175-184.

Locatelli, Claudriana, Rober Rosso, Maria C. Santos-Silva, Camila A. de Souza, Marley A. Licínio, Paulo Leal, Maria L. Bazzo, Rosendo A. Yunes, and Tânia B. Creczynski–Pasa. "Ester derivatives of gallic acid with potential toxicity toward L1210 leukemia cells." *Bioorganic & Medicinal Chemistry* 16, no. 7 (2008): 3791-3799.

Mukherjee, Gargi, and Rintu Banerjee. "Production of gallic acid. Biotechnological routes (Part 2)." *Chimica Oggi* 21, no. 3-4 (2003): 70-73.

Ortega, E., M. C. Sadaba, A. I. Ortiz, C. Cespon, A. Rocamora, J. M. Escolano, G. Roy, L. M. Villar, and P. Gonzalez-Porque. "Tumoricidal activity of lauryl gallate towards chemically induced skin tumors in mice." *British Journal of Cancer* 88, no. 6 (2003): 940-943.

Patil, P. and S. Killedar (2021). "Formulation and characterization of gallic acid and quercetin chitosan nanoparticles for sustained release in treating colorectal cancer." *Journal of Drug Delivery Science and Technology* 63: 102523.

Patil, P. and S. Killedar (2022). "Green Approach Towards Synthesis and Characterization of GMO/Chitosan Nanoparticles for In Vitro Release of Quercetin: Isolated from Peels of Pomegranate Fruit." *Journal of Pharmaceutical Innovation* 17(3): 764-777.

Rosmorduc, Olivier, and Christèle Desbois-Mouthon. "Targeting STAT3 in hepatocellular carcinoma: Sorafenib again…." *Journal of hepatology* 55, no. 5 (2011): 957-959.

Roy, Rituparna, Debolina Pal, Subhayan Sur, Suvra Mandal, Prosenjit Saha, and Chinmay Kumar Panda. "Pongapin and Karanjin, furanoflavanoids of Pongamia pinnata, induce G2/M arrest and apoptosis in cervical cancer cells by differential reactive oxygen species modulation, DNA damage, and nuclear factor kappa-light-chain-enhancer of activated B cell signaling." *Phytotherapy Research* 33, no. 4 (2019): 1084-1094.

Samad, N., and A. Javed. "Therapeutic effects of gallic acid: Current scenario." *J Phytochemistry Biochem* 2, no. 113 (2018): 2.

Sanchez-Martin, V., Plaza-Calonge, M. del C., Soriano-Lerma, A., Ortiz-Gonzalez, M., Linde-Rodriguez, A., Perez-Carrasco, V., Ramirez-Macias, I., Cuadros, M., Gutierrez-Fernandez, J., Murciano-Calles, J., Rodríguez-Manzaneque, J. C., Soriano, M., & Garcia-Salcedo, J. A. "Gallic Acid: A Natural Phenolic Compound Exerting Antitumoral Activities in Colorectal Cancer via Interaction with G-Quadruplexes." *Cancers* 14, no. 11 (2022): 2648.

Schulz, G. B., C. G. Stief, and B. Schlenker. "Follow-up surveillance of muscle-invasive urinary bladder cancer after curative treatment." *Der Urologe. Ausg. A* 58, no. 9 (2019): 1093-1106.

Shahid, Rizwana, Sadia Khan, Qaiser Aziz, and Muhammad Umar. "Frequency and diversity of the cases reported at Oncology care clinic of Holy Family Hospital Rawalpindi during August 2022: Frequency and Diversity of the Cases Reported at Oncology Care Clinic." *Pakistan Journal of Health Sciences* (2022): 101-104.

Singh, Brahma N., Braj R. Singh, B. K. Sarma, and H. B. Singh. "Potential chemoprevention of N-nitrosodiethylamine-induced hepatocarcinogenesis by polyphenolics from Acacia nilotica bark." *Chemico-biological interactions* 181, no. 1 (2009): 20-28.

Sivaraman Siveen, Kodappully, Kirti Prabhu, Roopesh Krishnankutty, Shilpa Kuttikrishnan, Magdalini Tsakou, Feras Q Alali, Said Dermime, Ramzi M Mohammad, and Shahab Uddin. "Vascular endothelial growth factor (VEGF) signaling in tumor vascularization: potential and challenges." *Current Vascular Pharmacology* 15, no. 4 (2017): 339-351.

Song, Bingbing, Jia Li, and Jianke Li. "Pomegranate peel extract polyphenols induced apoptosis in human hepatoma cells by mitochondrial pathway." *Food and Chemical Toxicology* 93 (2016): 158-166.

Sroka, Z., and W. Cisowski. "Hydrogen peroxide scavenging, antioxidant and anti-radical activity of some phenolic acids." *Food and Chemical Toxicology* 41, no. 6 (2003): 753-758.

Subramanian, A. P., A. A. John, M. V. Vellayappan, A. Balaji, S. K. Jaganathan, Eko Supriyanto, and Mustafa Yusof. "Gallic acid: prospects and molecular mechanisms of its anticancer activity." *Rsc Advances* 5, no. 45 (2015): 35608-35621.

Subramanian, Vimala, Balaji Venkatesan, Anusha Tumala, and Elangovan Vellaichamy. "Topical application of Gallic acid suppresses the 7, 12-DMBA/Croton oil induced two-step skin carcinogenesis by modulating anti-oxidants and MMP-2/MMP-9 in Swiss albino mice." *Food and Chemical Toxicology* 66 (2014): 44-55.

Tsai, Chung-Lin, Ying-Ming Chiu, Tin-Yun Ho, Chin-Tung Hsieh, Dong-Chen Shieh, Yi-Ju Lee, Gregory J. Tsay, and Yi-Ying Wu. "Gallic acid induces apoptosis in human gastric adenocarcinoma cells." *Anticancer Research* 38, no. 4 (2018): 2057-2067.

Wang, Ruixuan, Lijie Ma, Dan Weng, Jiahui Yao, Xueying Liu, and Faguang Jin. "Gallic acid induces apoptosis and enhances the anticancer effects of cisplatin in human small cell lung cancer H446 cell line via the ROS-dependent mitochondrial apoptotic pathway." *Oncology Reports* 35, no. 5 (2016): 3075-3083.

Wang, Yingqiang, Qianqian Luo, Youping Li, Shaolin Deng, Xianglian Li, and Shiyou Wei. "A systematic assessment of the quality of systematic reviews/meta-analyses in radiofrequency ablation versus hepatic resection for small hepatocellular carcinoma." *Journal of Evidence-Based Medicine* 7, no. 2 (2014): 103-120.

Wen, Haiyan, Siqi Zhou, and Jinchun Song. "Induction of apoptosis by magnolol via the mitochondrial pathway and cell cycle arrest in renal carcinoma cells." *Biochemical and Biophysical Research Communications* 508, no. 4 (2019): 1271-1278.

You, Bo Ra, and Woo Hyun Park. "Gallic acid-induced lung cancer cell death is related to glutathione depletion as well as reactive oxygen species increase." *Toxicology in Vitro* 24, no. 5 (2010): 1356-1362.

Zahrani, Nourah A. AL, Reda M. El-Shishtawy, and Abdullah M. Asiri. "Recent developments of gallic acid derivatives and their hybrids in medicinal chemistry: A review." *European Journal of Medicinal Chemistry* 204 (2020): 112609.

Zhao B, Hu M. Gallic acid reduces cell viability, proliferation, invasion and angiogenesis in human cervical cancer cells. *Oncology Letters*. 2013 Dec 1;6(6):1749-55.

Chapter 8

In Vitro Anticancer Activity Gallic Acid Nanoparticles on Colon Cancer Cell Colo 205

Poournima Sankpal[1,*], PharmD
Sachinkumar Patil[1], M.Pharm. PhD
Pramod B. Patil[1]
Rajanikant Ghotane[1]
Prafulla Choudhari[2], M.Pharm. PhD
and Sanket Rathod[2]

[1] Ashokrao Mane College of Pharmacy, Peth-Vadgaon, Kolhapur, Maharashtra, India
[2] Bharati Vidyapeeth College of Pharmacy, Kolhapur, Maharashtra, India

Abstract

The main objective of this study was to evaluate cytotoxic activity of gallic acid-loaded sodium alginate nanoparticles using GMO and poloxamer 407 toward human colorectal cancer cell lines as COLO 205. Gallic acid was successfully encapsulated into nanoparticles after being characterized with particle size entrapment effectiveness, FT-IR, X-ray diffraction, DSC, and loading content. The dialysis membrane approach was employed during *in vitro* drug release study at different pH levels (1.2, 4.5, 7.5, and 7.0) to mimic the GIT condition. After 24 hours 79.06% of drug release was accomplished in a sustained manner.

---

[*] Corresponding Author's Email: poournima6@gmail.com.

In: The Chemistry of Gallic Acid and Its Role in Health and Disease
Editor: Jeff C. Murdoch
ISBN: 979-8-88697-672-4
© 2023 Nova Science Publishers, Inc.

Results from the MTT assay on the human COLO 205 cell line showed that gallic acid nanoparticles had more significant anti-colon cancer activity, with an IC50 value of 6.99ug/ml. However, currently, no specific study on anti-cancer activity has been published on gallic acid nanoparticles by *in vitro* COLO 205 cell line model.

Keywords: gallic acid nanoparticles, cytotoxicity study, colon cancer

Introduction

It is well established that phenolics have potent antioxidant properties and can shield cells from oxidative damage brought on by free radicals [1, 2]. Due to their conjugated ring structures, numerous phenolic composites can function as superoxide species and shows the antioxidant effect by shielding superoxide anions and lipid peroxy radicals [3].

The phenolics are the primary chemical components of amla having potent antioxidant properties. There are several active substances, including chebulinate, quercetin, ellaginate, gallic acid, 1-O-galloyl-D-glucose, and chebulagic acid present in amla fruit [4]. Other active substances include kaempferol, mucic acid 1,4-lactone 3-O-gallate, isocorilagin, chebulanin, etc., also present in the amla fruit [5].

A naturally occurring phenolic substance (gallic acid) obtained from the fruits of the *Emblica Officinal*. However, due to extremely poor water solubility, significant presystemic anabolism, and catabolism, it shows reduced absorption and bioavailability in the body, which is a major obstacle to using it as a chemotherapeutic medication. It is also very sensitive to light and gets degrades quickly in both neutral and alkaline pH environments. Gallic acid has been formulated into a variety of forms, including microparticles, liposome, and solid dispersion agents to address these issues [6]. Polymeric NPs usually exhibit a sharper size and distribution, greater stability, greater harmonized physicochemical properties, sustained and regulated drug-release patterns, and a higher loading capacity for weakly water-soluble compounds. The surface characteristics of the SLN are rarely modified or coated to reduce the macrophage phagocytic uptake while improving the pharmacology of these colloidal carriers [7-9].

The Glyceryl Monooleate (GMO), a surface-modified nanoparticulate system called sodium alginate nanoparticles in which GMO as a lipid component and sodium alginate as the coating polymer with poloxamer 407 used as a stabilizer. The US FDA has authorized the stabilizer poloxamer 407, a non-ionic block copolymer comprising oxides of polypropylene and polyethylene, for use in oral, parenteral, ophthalmic, and cutaneous formulations [5, 6, 10].

To our knowledge, there is no research reported on the use of the high-pressure homogenization (HPH) for preparation of nanoparticles using GMO/sodium alginate system where phytoconstituent gallic acid isolated from amla fruit. There are different methods for formulation of nanoparticles which include high-pressure homogenization, solvent-diffused micro-emulsion, dual emulsion technique, solvent emulsification-evaporation, and ultrasound [9].

Because of some advantages, the HPH method has indeed been regarded as the most excellent method, including even particle size distribution, improved dispersion of preparations using high lipid contents, avoidance of or lower volume of organic solvents, acceptance by regulatory authorities of equilibration equipment, and viability of scaling up for large-scale production [10].

With gallic acid serving as a model hydrophobic phytoconstituent, studying *in vitro* cell line anti-cancer activity on the COLO 205 cell line's which is the main objective of formulating gallic acid into nanoparticles, by GMO/sodium alginate system [11].

Materials and Methods

Materials

Poloxamer 407, Sodium alginate, Glyceryl Mono Oleate, and standard gallic acid were purchased from Research Lab. Mumbai. We purchased all of the chemicals and solvents from Himedia Laboratories.

Plant Material
In the Kolhapur district, a sample of different amla plant parts was collected. The voucher herbarium (PSP-1) was deposited in the department of pharmacognosy, Ashokrao Mane College of Pharmacy, Peth-Vadgaon Kolhapur. Between Januaryary to march, amla fruits were harvested and dried in the air for 10 to 15 days in shade.

Soxhlet Extraction Method

Chloroform, ethanol, and ethyl acetate were among the several solvents that were used to accurately extract phenolics from plants. The amla fruit dried powder was extracted using 800 ml over six hours using several solvents. To calculate the % yield for each of the three different solvents, The filtrate was collected before being dried at 60°C under reduced pressure with a rotary evaporator [12].

Phytochemical Screening

Qualitative Test
The biomolecules included in the plant extract were identified using phytochemical analysis, which was utilized to find the presence of primary and secondary metabolites. Alkaloids, glycosides, saponins, tannins, triterpenoids, steroids, flavonoids, and carbohydrates are among the phytochemical components of amla extract that have been tested.

Determination of Total Phenolic Content from Amla Extract
Phytochemical such as phenolics which exhibit redox properties that control their antioxidant activity. A colorimetric technique called Folin-method Ciocalteu's (FC) relies on the transport of electrons between chemicals and polyphenols. For determining the phenolic component, various extracts of solvents chloroform, ethyl alcohol, and ethyl acetate extracts of amla fruit were utilized. The reaction mixture contains by mixing 1 ml of methanolic solution of all extracts, 2.5ml of 10% Folin-Ciocalteu's reagent, and 2.5ml of 7.5% $NaHCO_3$. The sample was further incubated in a thermostat for 45 minutes at 45°C. The standard gallic acid solution underwent the same treatment (Standard) in methanol (10 to100μg/ml) used for calibration curve, and the absorbance was calculated at λmax of 765nm [13].

Techniques of Isolation and Purification of Bioactive Molecules from Amla Extract [14, 15]

Fractionation of Bioactive Compound by Flash Chromatographic Technique
Analysis was performed using a flash chromatography system, having TBP2H02 pump, TBD2000 UV detector, and automatic fraction collector. Data monitoring throughout the analysis was done using a system using

Chromo station software. On an OROCHEM OROFLO-4SiHPS column containing silica particles, the separation was done.

Gas Chromatography (GC)
Gas chromatography was carried out on 7890 B with Agilent DB 624 column using helium gas at 1 ml/min flow mode reference solution tetrahydrofuran. At 50°C, GC temperature was set to 250°C at 20°C/min.

Thin Layer Chromatography (TLC)
By examining the Rf values of components in different solvent systems then the optimum solvent system for amla fruit extracts be determined. TLC profiles of amla fruit in various solvent systems reveal the existence of a variety of Phytochemical in the given amla fruit extract.

Structural Clarification of the Bioactive Molecules [16-18]

The flash chromatographic fraction of amla fruit extract given as number FA004 were filtered, dried, and stored at 4°C for further FT-IR and 1H-NMR analysis and HPC analysis.

UV-Visible Spectroscopy
Separated component from flash chromatographic was scanned and the experimental solution was examined with maximum wavelength at 270nm spectra.

FTIR Spectroscopy
FTIR has been demonstrated to be an effective method for the characterization and detection of functional groups that exist in plant extract compounds. Infrared spectra were noted using IR functioned form 4000–600cm^{-1} at a resolution of 4cm^{-1}. Opus software was used to analyze data.

NMR Spectroscopy of the Isolated Compound
Only fraction FA004 from was investigated further by 1HNMR utilizing solvent D6 + CDCL3 MIX. The analysis was performed on a 400 MHz BRUKER device.

HPLC of Isolated Compounds

HPLC PU-2080 Plus with UV-2075 plus intelligent detector and HPLC C18 column (250×4.6mm, 5μm) was set at 270nm for estimation of gallic acid. In the dynamic phase for compound elution, a 40:60 ratio of acetonitrile and 2% acetic acid was used. Gallic acid content was determined quantitatively in fraction concentrations (FA004) ranging from 0.01 to 0.5mg/ml.

HPTLC of the Isolated Compound

A Camag HPTLC system and TLC Scanner III with Win CATS version 1.4.0 software were used for investigation. For quantitative measurement of gallic acid from amla fruit, the best solvent system was identified to be ethyl acetate: methanol: toluene (8:2:1). Different concentrations of standard solution (40 to 240ng/spot were spotted on the HPTLC plate.

Determination of Solubility of the Isolated Compound

The separated compounds were evaluated for solubility for various solvents such as DMSO, ethanol, methanol, and acetone.

Melting Point Determination

To determine the identification and purity of a separated biomolecule, its melting point was determined in a thermionic device. The separated compound's recorded melting point was compared to the standard melting point of the relevant gallic acid.

Antioxidant Activity by DPPH Method [19]

2, 2-diphenyl-1-picryl-hydrazyl (DPPH) is a chemically stable free radical that is utilized to measure the antioxidant activity of different substances and the DPPH radical scavenging action of amla fruit and an isolated fraction determined by the following procedure [20]. Following the scavenging capacity was determined using equation number (1) different concentrations were combined (1ml, 0.004% w/v) and measured at 517nm.

$$\text{Scavenging activity (\%)} = \frac{(\Delta A517 \text{ of control} - \Delta A517 \text{ of sample})}{\Delta A517 \text{ of control}} \times 100 \quad (1)$$

Formulation of Herbal Nanoparticles [21]

A GMO/sodium alginate was employed to prepare o/w micro-emulsions of separated gallic acid. Isolated gallic acid (100mg) was dissolved into molten GMO (2g), and 12.5ml of 0.1% poloxamer 407 was introduced in it and sonicated at 18 W for 3 minutes in a probe sonicator. Dropwise 12.5ml of 2.4% sodium alginate solution was mixed with this emulsion with the help of a probe sonicator at 16W for four minutes. Lastly, this phase was subjected to twelve cycles of HPH at 15,000 pressure to produce the nanoemulsion, which again was subsequently lyophilized for 48 hours using 2% mannitol as a cryoprotectant. The central composite design was used to investigate the cumulative influence of two factors, each at two levels, as well as the possible nine herbal nanoparticle combinations.

Characterization of Herbal Nanoparticles

Particle Size and Zeta Potential [22]
A particle Size Analyzer was used to figure out the average particle size as well as the zeta potential of the herbal nanomaterials.

The nanosuspension was diluted using filtered ultra-pure water to determine particle size.

By FTIR Spectroscopy
By connecting an ATR accessory to an FTIR spectrometer, attenuated total reflection/Fourier transform infrared spectroscopy (ATR/FTIR) spectra were obtained at ambient temperature (Perkin Elmer, Spectrum 100).The detection of functional groups found in compounds derived from plants has shown to be a useful application of FTIR.

Differential Scanning Colorimetry [23]
Gallic acid, poloxamer 407, sodium alginate, physical mixtures, as well as gallic acid-loaded nanoparticles' were studied using DSC (2.0mg 0.2). A sample of 3-5mg was crimped at $10°Cmin^{-1}$ at a heating rate of 30 to 300°C in an aluminum pan.

In Vitro *Release Studies [10]*

To assess the release of nanoparticles at a specific pH and to ascertain drug release, herbal nanoparticles were tested in a variety of simulated fluids at various levels of pH.

Four milligrams of herbal nanoparticles were spread in a phosphate-buffered saline (PBS; pH = 2.0, 4.5, 6.8, 7.4) as a release medium in a dialysis membrane sac (MWMW cut-off 12 kDa; Sigma Aldrich) to simulate ileocolon conditions for 24hr.50mL of the release media were placed in a beaker, and the encased dialysis sac was submerged in it and the beaker was placed in an incubator that was shaken at 37°C with mild agitation using PBS and pH values of 2.0 for the first four hours, 4.5 for the next five to nine hours, and 6.8 for the final ten to thirteen hours and finally pH = 7.4 for fourteen to twenty-four hours. At predetermined intervals, 5ml of the supernatant was removed and tested for drug release using a UV spectrophotometer and gallic acid at 270nm.

In Vitro Anticancer Activity by MTT Assay [24]

Culture of cells bought from the American Type Culture Collection Colo-205 cell line (ATCC, Manassas, VA, USA). The cells were incubated at 37°C in a humid environment comprising 5% CO_2 while being cultured in DMEM containing 10% fetal bovine serum, 2mM L-glutamine, and 1% penicillin/streptomycin.

The cells were rinsed with Dulbecco's Phosphate-Buffered saline solution and separated from the flasks using trypsin/EDTA, then centrifuged at 1000rpm for 5min at 25°C.

Result and Discussion

Soxhlet Extraction Method [25]

The Soxhlet extraction technique was used to separate the amla fruit utilizing three different solvents: chloroform, ethanol, and ethyl acetate. Since ethyl acetate solvent produces the maximum yield (42.51%), it is employed for phenolic extraction.

Phytochemical Screening [13]

Qualitative Tests
Phytochemical screening of amla fruit extract results in a positive test for phenolics, alkaloids, tannins, and carbohydrates.

Total Phenolic Content [26]
The standard gallic acid calibration curve revealed a linear equation at y = 0.015x±0.399, R2 = 0.999, and the amounts of phenolics in chloroform, ethanol, and ethyl acetate were 25.73±0.21, 42.09±0.19, and 63.76±0.29mg GAE/g, respectively. Ethyl acetate produced higher yields when compared to other solvents; for this reason, it is a good solvent to use when extracting phenolics [27, 28].

Techniques of Isolation and Purification of Bioactive Molecules from Amla Fruit

Fractionation of Bioactive Compound by Flash Chromatographic Technique
With flow rates set at 4ml/min and the wavelength for amla fruit at 270 nm, the mobile phase employed was ethyl acetate: methanol 100:0 to 0:100. Further in 25gm silica gel column was loaded with an 8.0gm slurry (3g extract and 5g silica gel) (200-400 mesh size). Individually combining 3g of silica gel with 1g of amla extract powder, the mixture was thoroughly chopped into small pieces in a mortar and pestle. Further, samples of extract that had been appropriately blended were placed in sample holders. The linear gradient was used to separate the five fractions, with a peak tube volume of 14ml and a run time of 15 minutes. From the amla extract, various fractions with numbers FA001 to FA005 were separated and dried on a buchiroto evaporator (R-210 water bath B-491) for dryness.

It was discovered that fraction FA004 had a 33.4 mg/gm as a percentage yield. The UV spectra of fraction no. FA004 phytoconstituent, which produces absorbance at 270.5nm, comprises one of the five fractions of amla extract. This absorbance was validated by standard gallic acid solution spectra at 272nm. For better outcomes, these isolated fractions were also FA004 and analyzed by IR, H1NMR, HPLC, and HPTLC methods.

Separation by Using Column Chromatography

Simple fractional distillation was used to recover the solvent from isolated fractions no FA004 that were concentrated in vacuo. The mobile phase that best separated the compounds for the fraction further on TLC and was further described by IR, 1HNMR, and GCMS methods was composed of ethyl acetate: methanol: toluene (8:2:1).

Gas Chromatography

To estimate the amount of ethyl acetate in amla extract, the work demonstrated a straightforward gas chromatographic approach. The retention period is 4.526min according to the GC analysis of the amla crude ethyl acetate extraction. The amount of ethyl acetate detected in amla fruit was 1305.376ppm. Outstanding findings were found, especially when taking into account the low concentration levels examined, within the globally accepted validation reference standards.

Thin Layer Chromatography

The mobile phase, which was composed of ethyl acetate, methanol, and toluene (8:2:1), was the stage that best-separated compounds out of all those that were examined.

When compared to a standard control sample that had an Rf value of 0.86 and had been run in the same solvent system, the sample's Rf value was found to be 0.86.

To obtain a pure chemical, column chromatography was performed using the same mobile system.

Structural Clarification of the Bioactive Molecules

Utilizing FT-IR, 1H-NMR, and HPLC technology, the isolated chemicals (by flash chromatography) were analyzed and quantitatively evaluated.

UV-Visible Spectroscopy

UV/Vis spectrophotometer was used to find fraction FA004 and the specified sample solution's UV spectrum was screened, and the maximum wavelength at 270nm was recorded. Gallic acid showed good linearity from 2 to 18g/mL, with a linear regression equation of $y=0.059x\pm0.018$ and a correlation coefficient of $r^2=0.995$.

FTIR Spectroscopy of the Isolated Compound
Functional groups such as hydroxyl (-OH) stretch, C-H stretch for alkenes, C=O stretch for acid, and aromatic benzenoid ring were detected in the FT-IR spectra of isolation of amla fruit extract.

NMR Spectroscopy of the Isolated Compound
The investigation was carried out at the BRUKER device of 400 MHz d 9.136 (1H, H-7, s), 7.08 (1H, H-2, H-6, s), and 5.011 (1H, H-3, H-4, H-5, s). Using 1H NMR, isolated fraction of amla fruit revealed the existence of 7 carbon atoms and the molecular formula $C_7H_6O_5$, as well as aromatic, acidic, and hydroxyl protons.

HPLC Analysis of Isolated Compound
A comparative study was done with the peak of standard gallic acid (3.207min) and the spectrum of fruits of amla extract of different concentrations of 5 to 15g/mL of gallic acid, excellent linearity was observed; the linear regression equation was found to be y = 8008x-397.0 (R^2 = 0.999). By using HPLC the amount of gallic acid from the amla extracts, was found to be 27.15±0.001g/mg GAE equivalent [29].

Antioxidant Activity

Antioxidant Activity by DPPH Method [30]
The DPPH is a stable free radical that is extensively used to calculate the antioxidants' capacity to scavenge free radicals. The following procedure was used to measure the fruits' extract and isolated fraction's by DPPH radical-scavenging capacity. An aliquot of DPPH (1ml, 0.004% w/v) was combined with the extract at various concentrations and after vigorous shake; they were allowed to stand at room temperature for 30 minutes in the dark. The absorbance of this solution was measured at 517nm and the percentages of DPPH was calculated [31, 32].

Determination of Solubility of the Isolated Compound
The separated fraction was studied for solubility in various solvents. Isolated fraction is a white powder that can be dissolved with ether, ethanol, methanol, glycerol, and acetone.

Melting Point Determination

To identify and assess the purity of a substance, its melting point was measured using thermionic equipment. Comparing the measured melting point of the isolated compound of amla extract to the standard melting point gallic acid (260°C), the observed melting point was between 255-257°C.

Formulation of Herbal Nanoparticles [33]

The optimization aim of this work was to minimize particle size and maximize zeta potential. By factorial design demonstrating optimal condition for formulating herbal nanoparticles using 2.4% sodium alginate and poloxamer (407) 0.1% to obtain particle size 218.33 nm with zeta potential 11.50mV with desirability 1.000.

Characterization of Herbal Nanoparticles

Analysis of Particle Size and Zeta Potential

The average particle size of nanoparticles was found to be 214.21nm with a zeta potential of +14.7mV, due to the presence of protonated amino groups, has a positive charge in acidic solutions, making it suitable for adhering to the negatively charged gastrointestinal mucosal layer.

FTIR of Herbal Nanoparticles [34]

The spectrum of the characteristic peak formed by FTIR for the sodium alginate shows a stretch of carbonyl (ester) groups present in the sodium alginate, prominent bands may be seen in the spectra between 1150-1190cm^{-1} and 1240-1270cm^{-1}. At 1734.01cm^{-1}, There are also C (=O) ester vibration stretching bands. The carbon chain of poloxamer 407 showed at 2881.11 cm-1 for aliphatic C-H stretching, 1365.12 cm-1 and 1242.02cm-1 for planar O-H bend, 1096.99cm-1 for C-O stretch, and 840.46cm-1 for CH=CR2. A significant peak at 1738cm-1 was observed for the C=O functionality of GMO. A large band at 3194.61 cm-1 in the spectra of gallic acid is associated with OH stretching and hydrogen bonding amongst phenolic hydroxyl groups. At 1255.93cm-1, the COOH stretch/bend was seen. At 1454.44cm-1, aromatic ring stretching was observed. Gallic acid-loaded nanoparticles' spectra revealed that the O-H stretch had vanished. According to all available FTIR data, gallic acid was encapsulated into herbal nanoparticles with intermolecular hydrogen bonding in a nanoformulation that was less crystalline than a pure biomolecule.

Differential Scanning Calorimetry

For gallic acid, a prominent endothermic peak was detected at 259.68°C, which is also its melting point. Some other peak was seen at 91.73°C which may have been caused by water evaporation and revealed the crystal structure of pure gallic acid. When Poloxamer 407 melted, it produced a primary endothermic peak with a pointed peak at 58.92°C. While a significant endothermic peak was visible on the sodium alginate thermogram between 87.13°C and 226°C. The thermogram of the physical mixture exhibited no variations; rather, it just shows the temperature profiles of the separate components superimposed, showing that there was no interaction between them. On the other hand, the thermal analysis of sodium alginate nanoparticles loaded with gallic acid revealed the elimination of the drug's major peak, demonstrating total encapsulation of gallic acid within the polymeric nanoparticles [35].

In Vitro *Release Studies*

As an outcome, herbal nanoparticles increased drug release by 77.56% for gallic acid at 24 hours. As a result, herbal nanoparticles might be regarded as a possible barrier capable of releasing the biomolecule in colonic pH. Gallic acid biomolecule acquired controlled and sustained release through engineering sodium alginate method and benefited from its targeting capability to colonic region. The results were projected into a few kinetic models and intended specifically information into the Korsmeyer-Peppas power law model to explain the process of gallic acid discharge from the herbal nanoparticle. The Peppas (power low) models as well as zero order discharge case II transport (n = 2.09) were found to best fit the releasing of gallic acid from nanoparticles at pH 7.4 [10, 36].

In Vitro **Anticancer Activity by MTT Assay**

Colo 205 cells were subsequently cultivated in cell culture flasks until they reached the appropriate 80% confluence, at particular interval of time they were removed and seeded on E-plate for 30 minutes at room temperature. To determine the best seeding concentration, E-plate 96 was placed in the cell culture incubator with varied cell numbers. Because of the log growth phase attained on this cell number, we opted to seed 12500 cells/well for optimal seeding.

In addition to our findings on the Colo-205 cell line, our finding suggests that herbal nanoparticles containing gallic acid may be a promising alternative or adjunct to traditional neoplastic agent. It was critical for us to demonstrate when this cytotoxic impact begins and which dose would be the ideal treatment dose for the colon cancer cell line. Our results also validated real-time monitoring outcomes for gallic acid herbal nanoparticles. The IC50 value for gallic acid herbal nanoparticles was found to be 6.99g/mL. When compared to traditional endpoint cell-based assays, dynamic monitoring of cell response, such as cell adhesion, proliferation, and cell survival, is one of the benefits of the this system for and optimizing cell concentration for *in vitro* and *in vivo* assays, as well as allowing both cell and assay criteria to be acquired continuously before and during the experimentation [37]. The cytotoxic activity of herbal nanomaterial of gallic acid over human Colo-205 cell line was examined in this work employing an actual cell analyzer system for time-dependent observation.

Conclusion

Finally, the presence of phytoconstituent that is gallic acid in nanoformulation may be leading to the mechanism of cytotoxic activity against Colo-205 cell line shown in our in-vitro experiments. Though cytotoxic assay is sufficient for defining the mechanism of action of gallic acid in herbal nanoformulations, it is highly beneficial for giving a preliminary analysis that could be supported by a more precise bioassay.

Special findings related the colonic site included model hydrophobic biomolecule gallic acid along with nano size, positive charge on particle with good value, and sustained in-vitro releases at different P^H conditions. As a result, the discovery and creation of new nanoformulations based on natural products have been reported to have a regulated effect on the bioassay of the Colo-205 cell line; hence, they have the potential to be employed as a significant medicinal anticancer biomolecule. More research is needed to determine the likely exact mechanism through which gallic acid nanoparticles produce anticancer effects.

References

[1] Variya BC, Bakrania AK, Patel SS. Emblica officinalis (Amla): A review for its phytochemistry, ethnomedicinal uses and medicinal potentials with respect to molecular mechanisms. *Pharmacological Research,* 2016; 111:180-200. https://pubmed.ncbi.nlm.nih.gov/27320046/.

[2] Vadde R, Radhakrishnan S, Kurundu HEK, Reddivari L, Vanamala JK. Indian gooseberry (Emblica officinalis Gaertn.) suppresses cell proliferation and induces apoptosis in human colon cancer stem cells independent of p53 status via suppression of c-Myc and cyclin D1. *Journal of Functional Foods,* 2016; 25:267-78. https://pennstate.pure.elsevier.com/en/publications/indian-gooseberry-emblica-officinalis-gaertn-suppresses-cell-prol.

[3] Packirisamy RM, Bobby Z, Panneerselvam S, Koshy SM, Jacob SE. Metabolomic analysis and antioxidant effect of amla (Emblica officinalis) extract in preventing oxidative stress-induced red cell damage and plasma protein alterations: An *in vitro* study. *Journal of Medicinal Food,* 2018; 21(1):81-9. https://pubmed.ncbi.nlm.nih.gov/29064307/.

[4] Mastrodi Salgado J, Baroni Ferreira TR, de Oliveira Biazotto F, dos Santos Dias CT. Increased antioxidant content in juice enriched with dried extract of pomegranate (Punica granatum) peel. *Plant foods for Human Nutrition,* 2012; 67(1):39-43. https://pubmed.ncbi.nlm.nih.gov/22392496/.

[5] Patil P, Killedar S. Chitosan and glyceryl monooleate nanostructures containing gallic acid isolated from amla fruit: Targeted delivery system. *Heliyon,* 2021; 7(3):e06526. https://www.sciencedirect.com/science/article/pii/S2405844021006290.

[6] Patil P, Killedar S. Formulation and characterization of gallic acid and quercetin chitosan nanoparticles for sustained release in treating colorectal cancer. *Journal of Drug Delivery Science and Technology,* 2021; 63:102523. https://www.researchgate.net/publication/351037299_Formulation_and_Characterization_of_gallic_acid_and_quercetin_chitosan_nanoparticles_for_sustained_release_in_treating_Colorectal_Cancer.

[7] Mohammed MA, Syeda JT, Wasan KM, Wasan EK. An overview of chitosan nanoparticles and its application in non-parenteral drug delivery. *Pharmaceutics,* 2017; 9(4):53. https://pubmed.ncbi.nlm.nih.gov/29156634/.

[8] Sahle FF, Balzus B, Gerecke C, Kleuser B, Bodmeier R. Formulation and *in vitro* evaluation of polymeric enteric nanoparticles as dermal carriers with pH-dependent targeting potential. *European Journal of Pharmaceutical Sciences,* 2016; 92:98-109. https://pubmed.ncbi.nlm.nih.gov/27393341/.

[9] Ahmad N, Ahmad R, Ahmad FJ, Ahmad W, Alam MA, Amir M, Ali A. Poloxamer-chitosan-based Naringenin nanoformulation used in brain targeting for the treatment of cerebral ischemia. *Saudi Journal of Biological Sciences,* 2020; 27(1):500-17. https://pubmed.ncbi.nlm.nih.gov/31889876/.

[10] Patil P, Killedar S. Green Approach Towards Synthesis and Characterization of GMO/Chitosan Nanoparticles for In Vitro Release of Quercetin: Isolated from Peels

of Pomegranate Fruit. *Journal of Pharmaceutical Innovation,* 2022; 17(3):764-77. https://en.x-mol.com/paper/article/1386868543094546432.

[11] Giuliano E, Paolino D, Fresta M, Cosco D. Mucosal applications of poloxamer 407-based hydrogels: An overview. *Pharmaceutics,* 2018; 10(3):159. https://pubmed.ncbi.nlm.nih.gov/30213143/.

[12] Bimakr M, Rahman RA, Taip FS, Ganjloo A, Salleh LM, Selamat J, Hamid A, Zaidul ISM. Comparison of different extraction methods for the extraction of major bioactive flavonoid compounds from spearmint (Mentha spicata L.) leaves. *Food and Bioproducts Processing,* 2011; 89(1):67-72. https://daneshyari.com/article/preview/19235.pdf.

[13] Sánchez-Rangel JC, Benavides J, Heredia JB, Cisneros-Zevallos L, Jacobo-Velázquez DA. The Folin–Ciocalteu assay revisited: Improvement of its specificity for total phenolic content determination. *Analytical Methods,* 2013; 5(21):5990-9. https://pubs.rsc.org/en/content/articlelanding/2013/ay/c3ay41125g.

[14] Bajpai VK, Kang SC. Isolation and characterization of biologically active secondary metabolites from Metasequoia glyptostroboides Miki Ex Hu. *Journal of Food Safety,* 2011; 31(2):276-83. https://onlinelibrary.wiley.com/doi/abs/10.1111/j.1745-4565.2011.00298.x.

[15] Still WC, Kahn M, Mitra A. Rapid chromatographic technique for preparative separations with moderate resolution. *The Journal of Organic Chemistry,* 1978; 43(14):2923-5. https://pubs.acs.org/doi/10.1021/jo00408a041.

[16] Bansal V, Sharma A, Ghanshyam C, Singla M. Rapid HPLC method for determination of vitamin c, phenolic acids, hydroxycinnamic acid, and flavonoids in seasonal samples of emblica officinalis juice. *Journal of Liquid Chromatography and Related Technologies,* 2015; 38(5):619-24. https://www.researchgate.net/publication/273311194_Rapid_HPLC_Method_for_Determination_of_Vitamin_C_Phenolic_Acids_Hydroxycinnamic_Acid_and_Flavonoids_in_Seasonal_Samples_of_Emblica_officinalis_Juice.

[17] Sawant L, Prabhakar B, Pandita N. Quantitative HPLC analysis of ascorbic acid and gallic acid in Phyllanthus emblica. *J Anal Bioanal Tech,* 2010; 1(2) https://www.omicsonline.org/quantitative-hplc-analysis-of-ascorbic-acid-and-gallic-acid-in-phyllanthus-emblica-2155-9872.1000111.php?aid=97.

[18] Vijayalakshmi R and Ravindhran R. Comparative fingerprint and extraction yield of Diospyrus ferrea (willd.) Bakh. root with phenol compounds (gallic acid), as determined by uv–vis and ft–ir spectroscopy. *Asian Pacific Journal of Tropical Biomedicine,* 2012; 2(3):S1367-S71. https://www.google.com/search?q=%5B18%5D+Vijayalakshmi+R+and+Ravindhran+R.+Comparative+fingerprint+and+extraction+yield+of+Diospyros+ferrea+(wild.)+Bakh.+root+with+phenol+compounds+(gallic+acid),+as+determined+by+uv%E2%80%93vis+and+ft%E2%80%93ir+spectroscopy.+Asian+Pacific+Journal+of+Tropical+Biomedicine,+2012;+2(3):S1367-S71&spell=1&sa=X&ved=2ahUKEwi-rp6yosT8AhVRPewKHVIRAp8QBSgAegQICBAB&biw=1366&bih=590&dpr=1.

[19] Sarabandi K, Jafari SM, Mohammadi M, Akbarbaglu Z, Pezeshki A, Heshmati MK. Production of reconstitutable nanoliposomes loaded with flaxseed protein

hydrolysates: Stability and characterization. *Food Hydrocolloids*, 2019; 96:442-50. https://agris.fao.org/agris-search/search.do?recordID=US201900370081.

[20] Smirnoff N, Cumbes QJ. Hydroxyl radical scavenging activity of compatible solutes. *Phytochemistry,* 1989; 28(4):1057-60. https://www.scirp.org/(S(i43dyn45 teexjx455qlt3d2q))/reference/ReferencesPapers.aspx?ReferenceID=1474075.

[21] Pandit AA, Dash AK. Surface-modified solid lipid nanoparticulate formulation for ifosfamide: Development and characterization. *Nanomedicine*, 2011; 6(8):1397-412. https://pubmed.ncbi.nlm.nih.gov/22091968/.

[22] Zhao GD, Sun R, Ni SL, Xia Q. Development and characterisation of a novel chitosan-coated antioxidant liposome containing both coenzyme Q10 and alpha-lipoic acid. *Journal of Microencapsulation*, 2015; 32(2):157-65. https://pubmed.ncbi.nlm.nih.gov/25329530/.

[23] Şanlı O, Kahraman A, Kondolot Solak E, Olukman M. Preparation of magnetite-chitosan/methylcellulose nanospheres by entrapment and adsorption techniques for targeting the anti-cancer drug 5-fluorouracil. *Artificial Cells, Nanomedicine, and Biotechnology*, 2016;44(3):950-9. https://www.tandfonline.com/doi/full/10.3109/21691401.2015.1008502.

[24] Patil P, Killedar S, More H, Vambhurkar G. Development and Characterization of 5-Fluorouracil Solid Lipid Nanoparticles for Treatment of Colorectal Cancer. *Journal of Pharmaceutical Innovation*, 2022:1-14.

[25] Altemimi A, Watson DG, Kinsel M, Lightfoot DA. Simultaneous extraction, optimization, and analysis of flavonoids and polyphenols from peach and pumpkin extracts using a TLC-densitometric method. *Chemistry Central Journal*, 2015; 9(1):1-15. https://pubmed.ncbi.nlm.nih.gov/26106445/.

[26] Ali G, Hawa ZJ, Asmah R. Effects of solvent type on phenolics and flavonoids content and antioxidant activities in two varieties of young ginger (Zingiber officinale Roscoe) extracts. *Journal of Medicinal Plants Research*, 2011; 5(7):1147-54. https://academicjournals.org/article/article1380532263_Ghasemzadeh%20 et%20al.pdf.

[27] Abdel-Aal EI, Haroon AM, Mofeed J. Successive solvent extraction and GC–MS analysis for the evaluation of the phytochemical constituents of the filamentous green alga Spirogyra longata. *The Egyptian Journal of Aquatic Research*, 2015; 41(3):233-46. https://www.sciencedirect.com/science/article/pii/S1687428515000515.

[28] Ghasemzadeh A, Jaafar HZ, Juraimi AS, Tayebi-Meigooni A. Comparative evaluation of different extraction techniques and solvents for the assay of phytochemicals and antioxidant activity of hashemi rice bran. *Molecules*, 2015; 20(6):10822-38. https://pubmed.ncbi.nlm.nih.gov/26111171/.

[29] Mangold H, Stahl E. *Thin-layer Chromatography*, 1969. https://link.springer.com/book/10.1007/978-3-642-88488-7.

[30] Alves DB. French reaction to the menace from Cabanos and Bonis within the litigious territory between Brazil and French Guiana (1836-1841). *Almanack*, 2016:126-95. https://www.scielo.br/j/alm/a/v5gDpzvYTDVPZZKRjPM9JNd/?lang=en.

[31] Luo W, Zhao M, Yang B, Shen G, Rao G. Identification of bioactive compounds in Phyllenthus emblica L. fruit and their free radical scavenging activities. *Food Chemistry,* 2009; 114(2):499-504. https://www.researchgate.net/publication/2383 78882_Identification_of_bioactive_compounds_in_Phyllenthus_emblica_L_fruit_ and_their_free_radical_scavenging_activities.

[32] Nawaz H, Shad MA, Rehman N, Andaleeb H, Ullah N. Effect of solvent polarity on extraction yield and antioxidant properties of phytochemicals from bean (Phaseolus vulgaris) seeds. *Brazilian Journal of Pharmaceutical Sciences,* 2020; 56. https://www.researchgate.net/publication/339979147_Effect_of_solvent_polarity_ on_extraction_yield_and_antioxidant_properties_of_phytochemicals_from_bean_ Phaseolus_vulgaris_seeds.

[33] Sun X, Wang Z, Kadouh H, Zhou K. The antimicrobial, mechanical, physical and structural properties of chitosan–gallic acid films. *LWT-Food Science and Technology,* 2014; 57(1):83-9. https://core.ac.uk/download/pdf/56685051.pdf.

[34] Itoh N, Katsube Y, Yamamoto K, Nakajima N, Yoshida K. Laccase-catalyzed conversion of green tea catechins in the presence of gallic acid to epitheaflagallin and epitheaflagallin 3-O-gallate. *Tetrahedron,* 2007; 63(38):9488-92. https://www.i nfona.pl/resource/bwmeta1.element.elsevier-7f63dc97-22d5-3479-83b4-e4a1d7b3 9b8d.

[35] Abdou EM, Masoud MM. Gallic acid–PAMAM and gallic acid–phospholipid conjugates, physicochemical characterization and *in vivo* evaluation. *Pharmaceutical Development and Technology,* 2018; 23(1):55-66. https://pubmed. ncbi.nlm.nih.gov/28627282/.

[36] Pasparakis G, Bouropoulos N. Swelling studies and *in vitro* release of verapamil from calcium alginate and calcium alginate–chitosan beads. *International Journal of Pharmaceutics,* 2006; 323(1-2):34-42. https://pubmed.ncbi.nlm.nih.gov/16828 245/.

[37] Lamarra J, Rivero S, Pinotti A. Design of chitosan-based nanoparticles functionalized with gallic acid. *Materials Science and Engineering: C,* 2016; 67:717-26. https://pubmed.ncbi.nlm.nih.gov/27287172/.

Chapter 9

Pharmacognosy of Gallic Acid and Its Co-Crystals

Sanchay Jyoti Bora[1,*], PhD
Riju Kakati Sarma[2], PhD
and Purabi Sarmah[3], PhD

[1] Department of Chemistry,
Pandu College, Pandu, Guwahati, Assam, India
[2] Department of Botany,
Pandu College, Pandu, Guwahati, Assam, India
[3] Department of Chemistry,
Nalbari College, Nalbari, Assam, India

Abstract

Gallic acid (GA), a naturally occurring low molecular weight polyphenolic compound, is known to possess tremendous health benefits. Its presence as a phytochemical in numerous plant species is of great biomedical significance. Studies have established the presence of GA in the fruit and fruit pods of *Caesalpinia coriaria*, *C. spinosa*, *Caesalpinia sappan*, *C. brevifolia* (Fabaceae); leaves of *Rhus typhina*, *R. coriaria* and leaf galls of *Rhus semialata* (Anacardiaceae); wood and bark of *Quercus* sp and *Castanea* sp (Fagaceae), wood galls of *Quercus infectoria* (Fagaceae); fruit of *Terminalia chebula* (Combretaceae) and leaves of *Ximenia americana* (Olacaceae) to name a few.

The fungal species *Aspergillus fischerii* MTCC 150 has been used to yield GA by the hydrolysis of tannic acid. Also, the bacterial species *Klebsiella pneumoniae* and *Corynebacterium* sp have been reported to

---

[*]Corresponding Author's Email: sanchay.bora@gmail.com.

In: The Chemistry of Gallic Acid and Its Role in Health and Disease
Editor: Jeff C. Murdoch
ISBN: 979-8-88697-672-4
© 2023 Nova Science Publishers, Inc.

produce it from the crude extract of tara gallotanin. This biologically relevant molecule (GA) has been found to possess various pharmacological activities like antioxidant, antibacterial, antitumor, antiviral properties etc. Besides these, GA has no toxicity or side effects even in large doses. Consequently, it has received much attention in the field of pharmacological research. This review presents a report on the natural occurrence of GA and its therapeutic properties on the basis of available literature.

Due to the presence of potential hydrogen bond donors and acceptor sites, GA appears to be an excellent candidate for co-crystal formation. On the other hand, low thermal stability, large particle size and poor solubility during absorption are some of the issues due to which pharmacological activities of GA are significantly diminished. The therapeutic effectiveness of these materials greatly depends on their solubility, because poor solubility can cause low bioavailability of the same. Literature reports reveal that the solubility and dissolution rate of GA can be greatly enhanced by constructing GA-based co-crystals with the help of several co-crystal formers (CCFs). In recent times, an increasingly larger section of chemists have started looking into the structures of binary crystalline solids with a view of examining the enhanced pharmacological properties in them. A consolidated account of reports available in literature related to the co-crystals comprised of GA, their formation, structure and pharmacological activities has also been provided in this chapter.

Keywords: gallic acid, pharmacognosy, pharmaceutical co-crystals

Introduction

The term pharmacognosy has been derived from the Greek words *'pharmakon'* meaning drug and *'gnosis'* meaning knowledge. It is the science of drugs of natural origin. It was first introduced by the Austrian physican Schmidt, in the year 1811, followed by Anotheus Seydler in 1815, in a work entitled *'Analecta Pharmacognostica'* (Tyler et al., 1988; Orhan, 2014).Towards the end of the 19th century and the beginning of the 20th century, the term 'pharmacognosy' got a new definition. It came to be used to define the branch of science dealing with medicines obtained from crude extracts of biological or mineral origin. At this stage the study was mostly concentrated on drugs of plant origin, both in their crude and powder forms.

The advent of the 21st century witnessed tremendous development in scientific know-how and ushered in a renaissance in pharmacognosical knowledge. The subject broadened to include studies at the molecular and metabolomic level (Orhan, 2014).

Mankind has been using plant products and its extracts as sources of food, medicine and numerous other products affecting their daily life, for centuries. However, the present-day scenario is slightly different. With an increase in scientific knowledge about food constituents and an unprecedented growth in health awareness among the masses, a growth in the demand for health promoting substances obtained through food sources, has been observed. One among the various health promoting substances are the antioxidants. They are a group of well-known biochemical compounds that have been widely studied for their tremendous potential in scavenging free radicals. The demand and search for antioxidants has resulted in the isolation of various compounds having antioxidant properties (Pandey and Rizvi, 2009). According to the Cambridge Dictionary, an antioxidant is "a substance that slows down the rate at which something decays because of oxidation: a chemical substance that prevents or slows down the damage that oxygen does to organisms or to food." The Oxford Dictionary defines it as "a substance such as Vitamin C or E that removes dangerous molecules, etc., such as free radicals from the body".

Halliwell et al., (1995) have defined antioxidants as substances that significantly delay or prevent the beginning or the propagation of oxidation chain reactions, at small concentrations as compared to the oxidizable substrate. In addition to the search for health promoting substances, there has been a growing realization about the health benefits, like tremendous antioxidant potential, of phytochemical compounds found abundantly in fruits (Basu and Penugonda, 2009), vegetables and other plant parts. Various phytochemicals like phenolic compounds, carotenoids, isoflavones etc. have been isolated from various plant parts. Two main groups of polyphenols are the flavonoids and non-flavonoid polyphenols (de la Rosa et al., 2010). The flavonoid group with a C_6-C_3-C_6 structure, is classified into three groups, namely – bioflavonoids, iso-flavonoids (phytoestrogens) and neo-flavonoids (white flavonoids).On the other hand, the non-flavonoids include phenolic acids, stilbenes and lignans. The phenolic acids have a carboxyl group attached or linked to a benzene ring (Lafay and Gil-Izquierdo, 2008). Two classes of phenolic acids can be distinguished depending on their structure: benzoic acid derivatives (i.e., hydroxybenzoic acids, C_6-C_1) and cinnamic acid derivatives (i.e., hydroxycinnamic acids, C_6-C_3). The phenolic acids like ferulic, caffeic, synergic and ellagic acids etc. present in plants are a diverse group broadly classified as hydroxybenzoic and hydroxycinnamic acids, that are derivatives of benzoic acid and cinnamic acid respectively (Soong and Barlow, 2004; Khadem and Marles, 2010). They are of great interest in food, cosmetic and pharmaceutical industries and also as substitutes for synthetic antioxidants. These phenolic phytochemicals constitute the most widely distributed group of phytochemicals in the plant kingdom. Almost all of the plants known to us have one or the other medicinal/health promoting constituents, known as phytoconstituents. The most important among them are the alkaloids, glycosides, tannins, flavonoids (How et al., 2008) and phenolic compounds (Hill, 1952). Among the dietary phenolics, the flavonoids, phenolic acids and polyphenols are the most important groups.

Gallic acid is one such naturally abundant polyphenolic phytochemical. Numerous research works conducted on gallic acid, have established its antioxidant, antimicrobial, anti-tumour and anticancer properties (Faried et al., 2007; Kaur et al., 2009; You et al., 2010; Borges et al., 2012; Olmedo-Juárez et al., 2019; Liang et al., 2022). This compound is a trihydoxybenzoic acid, frequently occurring in the form of ester free acids, hydrolysable tannins and catechin derivatives (Sneha et al., 2015) and is now known to be a potent free radical scavenger. Its scavenging properties makes it a much-sought-for phytochemical in relation to improvement of human health, especially the diseases related to oxidative stress (Thompson and Collins, 2013). Recent research and epidemiological works have established the therapeutic nature of gallic acid in the prevention and cure of neurodegenerative diseases (Lu et al., 2006; Korani et al., 2014), along with anti-inflammatory properties against chronic disorders. Gallic acid has also been found to possess the potential to reverse carcinogenesis, without affecting the healthy cells. Moreover, this awesome compound has also been reported to have preventive and therapeutic potential like antioxidant property, cytoprotective, neuroprotective, cardio-protective and vasoprotective (Youdim and Joseph, 2001; Karamae et al., 2005; Kaur et al., 2005; Nayeem et al., 2016; Singh et al., 2018). Gallic acid has also been reported to exhibit anti-cancer properties in prostate carcinoma cells (Chen et al., 2009; Kaur et al., 2009).

Another well-known group of phytochemicals, the tannins, which are polyphenolic secondary metabolites found in higher plants, have been defined as: "C- and O-glycosidic derivatives of gallic acid (3, 4, 5-trihydroxybenzoic acid)". However, this definition includes only some tannins, leaving out the condensed tannins containing the flavan-3-ol (catechin) units. The C- and O-derivatives have however been found to be of relatively low occurrence in its structure or sometimes the gallyol or its derivatives may be totally absent (Khanbabaee and Ree, 2001).

Phytosources of Gallic Acid

Numerous research works support the presence of Gallic acid,3, 4, 5-trihydroxybenzoic acid in a number of higher plant species. But lower plant groups like algae have not as yet been found to contain any related polyphenolic natural products. However, a number of Aspergillus species like *A.niger, A. awamori* (Pourrat et al., 1985; Seth and Chand, 2000) and *A. fischeri*, a hitherto unreported species (Bajpai and Patil, 2008) have been found to yield gallic acid by the hydrolysis of tannic acid. In addition to this, certain bacterial species like *Klebsiella pneumonia* and *Corynebacterium* sp. have been found to utilize crude extract of tara gallotannin to produce gallic acid (Deschamps et al., 1984).

Gallic acid and its derivatives are a group of phytochemicals occurring naturally in plants in the form of free acids, esters, catechin derivatives, hydrolysable tannins, etc.

Table 1 shows the common phytosources of gallic acid and its derivatives (Manach et al., Dalla et al., 2005; Liu et al., 2006; Chen et al., 2009; Aher et al., 2010; Shahriar and Robin, 2010; Baiano, 2013).

Research conducted on ayurvedic formulations has also proved the presence and therapeutic role of gallic acid and its derivatives (Kaur et al., 2005; Sandhya et al. 2006; Patel et al., 2010; Borde et al. 2011).

Table 1. The common phytosources of gallic acid and its derivatives

Scientific name	English name	Parts used
Allanblackia floribunda	Tallow-tree	Fruits and leaves
Allium cepa	Onion	Fleshy scale lvs/green lvs
Ananas comosus	Pineapple	Fruit
Anacardium occidentale	Cashew apples	Juicy apple-like thalamus
Arbutus unedo	Strawberry-tree	Fruit
Atraphaxis frutescens	Shrubby buckwheat	Aerial parts
Bergia suffruticosa	Woody bergia	Whole plant
Bridelia micrantha	Mitzeri/CoastalGolden-leaf tree	Stem bark
Caesalpinia sappan	Indian Redwoood/Sappanwood	Plant (heartwood) extract
Camellia sinensis	Tea	Leaves, bark
Carya illinoinensis	Mexican pecan	Kernal and shells
Casuarina equisetifolia	Coastal/Horsetail She-Oak	Wood extracts, bark, fruit, leaf

Table 1. (Continued)

Scientific name	English name	Parts used
Ceratonia siliqua	Carob/Locust bean	Pods/oil
Coffea arabica	Coffee	Bean
Cucumis sativus	Cucumber	Entire fruit /Rind/leaves
Citrus	Lemon	Entire fruit
Dillenia indica	Elephant apple	Entire fruit
Diospyros cinnabarina	Ebony/Persimmon	Entire fruit
Elephantorrhiza elephantine	Elephant's root/Eland's wattle	Roots/leaves
Eucalyptus gunnii	Cider gum	Leaves
Euphoria longana	Longan	Seed
Fragaria sp	Strawberry	Berry
Frankenia laevis	Sea-heath	Whole plant extract
Garcinia densivenia	Sap tree- mangosteen relative	Leaves and stem bark
Geranium collinum	Hill geranium	Root extracts
Haematoxylum campechianum	Logwood/Bloodwood	Stem bark
Hamamelis virginiana	American Witch-hazel	Leaves
Juglans sp	Walnuts	Nut (Kernel)
Juglans regia	English walnut	Nut (Kernel)
Mangifera indica L.	Mango	Seed kernel
Mimosa hamata	Hooked mimosa	Plant extract
Mimosa rubicaulis	Himalayan mimosa	Plant extract
Musa paradisica	Banana	Fruit, Peel
Oenothera biennis	Common evening primrose	Seed extract
Paeonia suffruticosa	Tree peony	Root bark
Paratecoma peroba	White peroba	Wood
Phyllanthus emblica	Indian gooseberry	Dry/Fresh Fruit -
Pista ciavera	Pistach walnut	Nut
Polygonum bistorta	Common bistort/snakeweed	Plant and rhizome extract
Poupartia axillaris	Chinese big head tea/ Fried egg plant	Wood
Psidium guajava	Guava	Fruit, leaves
Punica granatum	Pomegranate	Fruit rind and root bark
Pyrus communis	Pear	Fruit
Pyrus malus/ Malus pumila	Apple	Peels
Rheum spp	Rhubarb	Fleshy edible stalks
Rhus glabra	Smooth sumac	Root, leaf and seed extract
Rhus typhina	Staghorn sumac	Root, leaf and seed extract
Rubus idaeus	Red raspberries	Berries
Rubus laciniatus	Evergreen blackberries	Berries
Rubus occidentalis	Black raspberries	Berries
Rubus suavissimus	Chinese sweet leaf tea	Leaves and bud
Rubus ursinus	Marionberries	Berries
Sanguisorba officinalis	Great burnet	Underground parts

Scientific name	English name	Parts used
Satakentia liukiuensis	Palm tree	Fruit
Syzygium cordatum	Water-berry	Leaves, fruits
Tabernaemontana cymosa	Pinwheel flower/ Crape jasmine	Flower
Tamarix amplexicaulis	Tamarisk	Fruits
Tamarix nilotica	Nile Tamarix	Roots
Tectona grandis	Teak	Wood, bark, leaves
Terminalia bellarica	Beleric/Bastard myrobalan	Dry/Fresh fruit
Terminalia chebula	Black/chebulicmyrobalan	Dry/Fresh fruit
Toona sinensis	Chinese mahogany/Red toon	Leaf extract
Vitis rotundifolia	Muscadine grape	Dry/Fresh fruit
Vitis vinifera	Grape	Dry/Fresh fruit

Table 2. Ayurvedic formulations, their constituent plant species and medicinal properties

Ayurvedic formulation/ Name	Scientific name	English/Local name of plant used	Medicinal properties
Adulsa	*Adhatoda vasica*	Malabar nut	Treatment of bronchitis, tuberculosis
Aloe	*Aloe indica/Aloe vera*	Aloe	Antioxidant, antimicrobial
Anantmul/ Arkapani	*Tylophora asthmatica*	Country Ipecacuanha/ Indian Ipecac	Leaves used in treatment of bronchial asthma
Babool	*Acacia arabica*	Gum arabic	Bark- Anti diabetic, cure Haemorrhagic diseases Acacia Arabica (Babool) Diabetes (bark), Haemorrhagic diseases, Styptic.
Jatamansi	*Nardostachys jatamansi*	Spikenard	Treats epilepsy, hysteria, mental illness
Jaswand	*Hibiscus rosa-sinensis*	Hibiscus/China Rose	Reduces Hypertension; Chemopreventive effects
Jyeshthamadh (Mulethi)	*Liquorice glycerriza*	Licorice root	Reducing mouth and stomach ulcers
Nagarmotha	*Cyperus scariosus*	Nut grass	Reduces obesity, indigestion
Papaya	*Carica papaya*	Papaya/papaw	Antioxidant, Anticancer, anti-inflammation, protects skin, improves digestion
Pippali	*Piper longum*	Long pepper	Decongestant, bronchodilator in coughs and colds
Kutaki	*Picrorhiza kurroda*	Kutaki	Protects liver (used to cure Jaundice), antioxidant
Bakul	*Mimusops elengi*	Spanish cherry	Astringent, anthelmintic, tonic, febrifuge, alleviates *kapha* and *pitta*

Table 2. (Continued)

Ayurvedic formulation/ Name	Scientific name	English/Local name of plant used	Medicinal properties
Jamun	*Eugenia jambolana*	Malabar/Java/ Black Plum	Seed kernal aqueous extract: Anti-diabetic; Bark, root, seed, leaves- Astringent Leaves- Antibacterial
Raktachandan	*Pterocarpus santalinus*	Red sandalwood	Cooling property used for skin care products
Khair	*Acacia catechu*	Black katechu/ Catechu	Wood extract-Antibacterial, Antifungal, Anti-inflammatory, Antioxidant; Bark-Astringent, Anticancer
Behda	*Terminalia belerica*	Beleric or Bastard myrobalan	Fruit: Astringent, antibacterial
Shankhapushpi	*Convolvulus pluricalis/Clitoria ternetea*	Aloe weed/Aparajta	Stress and anxiety release, brain tonic
Manjishtha	*Rubia cordifolia*	Indian maddar	Blood purifier
Triphala	*Phyllanthus emblica*(Amla)+ *Terminalia chebula*(Hirda)+ *Terminalia belerica*(Behda)	Combination of three fruits: Amla, Hirda and Behada	Antibacterial, antiviral, antifungal, anti-inflammatory, antioxidant, antitumor
Amla	*Phyllanthus emblica*	Indian gooseberry	Dried Raw fruit: Anti-cancer, antibacterial, antiviral, astringent, anti-haemorrhagic, antidiabetic, antioxidant
Chyawanprash 12 ingredients: Amla, Neem, Pippali, Ashwagandha, Safed chandan, tulsi, elaichi, arjun,brahmi, kesar, ghrita (Desi ghee), shahad(Honey)	12 ingredients: *Phyllanthus emblica, Azadirachta indica, Piper longum, Withania somnifera, Santalum album, Ocimum sanctum, Elettaria cardamomum, Treminalia arjuna, Bacopa monnieri, Crocus sativus*	12 ingredients: Indian gooseberry, Neem, Long piper, Indian ginseng, White Sandalwood, Sacred Basil, Cardamom, Arjun, Indian pennywort, Saffron, Ghee, Honey	Antibacterial, antioxidant and all properties of the three constituents.
Neem	*Azadirachta indica*	Neem	Inflammation, skin, dental problems
Shallaki	*Boswellia serrata*	Indian Frankincense	Antiseptic, astringent, anti-arthritic
Lasuna	*Allium sativum*	Garlic	Reduces blood pressure, cholesterol, prevents cancer, antibiotic
Pudina	*Mentha arvensis*	Mint	Cleans toxins from the gut, strengthens stomach

Ayurvedic formulation/ Name	Scientific name	English/Local name of plant used	Medicinal properties
Dalimb	*Punica granatum*	Pomegranate	Fruit rind and root bark: Antibacterial, astringent
Lodhra	*Symplocos racemosa*	Symplocos bark/Lodh tree	Managing eye diseases, ulcers, haemorrhages
Hirda	*Terminalia chebula*	Chebulic myrobalan	Fruit: Antidiabetic, antibacterial, antioxidant, antitumor, astringent, may inhibit HIV
Kulanjana/Kulinjan	*Alpinia galanga*	Greater galangal	Antioxidant, anticancer, antimicrobial

Pharmacological Activities of Gallic Acid

Oxidative stress is the major cause of many degenerative diseases that occur in our body (Nandi et al. 2019). Accumulation and excessive production of free radicals formed in various metabolic processes can be attributed to the generation of oxidative stress. Reactive oxygen species (ROS) such as superoxide radical ($O_2^{\cdot -}$), hydroxyl radical ($OH^{\cdot}$), peroxy radicals($ROO^{\cdot}$), hydrogen peroxide(H_2O_2) etc. are often formed inside our body due to certain unknown redox system imbalances, influenced by a variety of internal and external factors. These ROS are actually responsible for causing cancer, cardiovascular diseases, aging, diabetes mellitus, neuro-disorders, etc.

Our body has some inherent antioxidant defense mechanisms to get rid of the deleterious effects produced by these reactive oxygen species. Some enzymatic antioxidants such as superoxide dismutase, catalase, glutathione peroxidase, glutathione reductase, glutathione-s-transferase etc. as well as some non-enzymatic antioxidants like polyamine, glutathione, bilirubin etc. are present in our body acting as free radical scavengers which can manage the concentration of reactive oxygen species up to a certain limit (Badhani et al. 2015 & Xu et al., 2021).

Some natural biomolecules such as vitamin C, vitamin E, carotenoids, flavonoids, polyphenolic compounds, anthocyanins etc. derived from a variety of food sources can also act as free radical scavengers and play an important role as external antioxidant defense agents. Gallic Acid (GA) is an important polyphenolic compound which exhibits the highest antioxidant properties among various other polyphenols because it has a greater number of hydroxyl groups attached to the aromatic ring in a position ortho to each other. The captodative effect played by the electron withdrawing group (EWG) and the electron donating groups (EDG) in GA has an influential role in stabilizing the free radical (B. Badhani et al. 2015). GA is found to have potential antioxidant properties with very low adverse effects. It is also found to be a very good apoptotic agent. The pharmacological effects of GA both *in vivo* and *in vitro* have been widely studied in recent times (N. Kahkeshani et al., 2019).

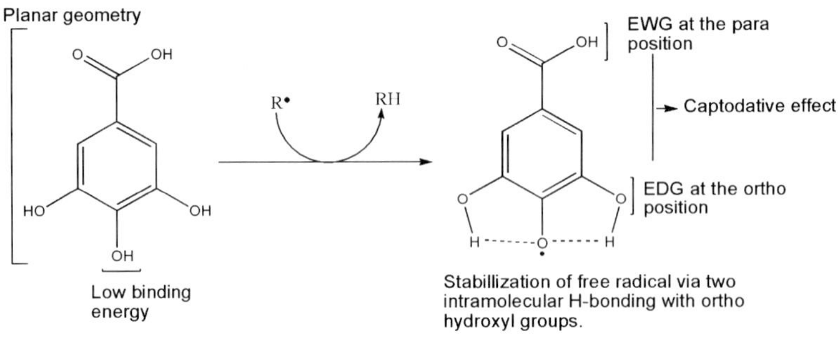

Figure 1. Stabilization of free radical in GA due to the captodative effect.

GA exhibits a promising neuroprotective effect against oxidative stress and brain cell damage (Shabani et al., 2020).It can be used as a therapeutic agent in controlling neurodegenerative diseases *viz.* Alzheimer's disease (AD), Parkinson's disease, depression, anxiety etc. The underlying cause of neurodegenerative disease can be ascribed primarily to oxidative stress, protein aggregation and cell death. People with AD are found to have aggregation of Amyloid beta (Aβ) protein called Amyloid plaques in the grey matter of their brain cells. Aβ proteins are formed due to some abnormal processing of the amyloid precursor protein (APP) which may produce oxidative stress and inflammatory response leading to a disorder of the nervous system. Metal ions are actually responsible for aggregation of proteins, which may interfere with the correct order of protein folding, leading to protein misfolding diseases. Cu (II) and Zn (II) metal ions can play an influential role in the formation of Amyloid plaques (Tougue et al., 2011). GA has the ability to form stable metal chelates thereby inhibiting the possibility of protein misfolding and metal induced aggregation (Khan et al., 2019).

GA is also found to play a therapeutic role against chemical induced neural dysfunctions. Due to massive urbanization and industrial developments, human beings are constantly exposed to harmful pollutants, which may cause neurotoxic effects in the long run such as behavioral changes, learning impairment, cognitive disorder, movement disorders, autism etc. GA and its derivatives are proven to be beneficial in mitigating some chemical induced neurotoxic effects. They can play an inhibitory role against oxidative stress and neurotoxicity by increasing the level of glutathione and also by increasing the activities of anti-oxidative enzymes like superoxide dismutase and catalase in different parts of our brain cells (Shabani et al., 2020).

GA can also exert promising anti-cancer activities by affecting various biological pathways such as metastasis, angiogenesis, migration, apoptosis, oncogene expression, etc. It can be used as a therapeutic drug in the form of nutritional supplement or in combination with vitamins against various types of cancers viz. lung cancer, colon cancer, stomach cancer, cervical cancer etc. due to its highly effective apoptotic property. Anti-cancer activity of GA is exerted due to its selective cytotoxicity against various cancerous cell lines viz. KATO III (stomach cancer), COLO 205 (colon cancer), A459 (lung cancer), HeLa (cervical cancer) etc. (Verma et al., 2013).

GA is also known to exhibit anti-allergic properties and is used in the treatment of allergic diseases like asthma, sinusitis, rhinitis etc. Allergies mostly occur in our bodies due to sudden release of inflammatory mediators such as histamine, heparin, various cytokines etc. stored in the granules of mast cells. Mast cells are actually some allergy cells of our body which can produce immediate response to different allergens, serving as the first line of defense in our body. Gallic acid can act as mast cell stabilizer and is effective against allergy antibody IgE-mediated mast cell degranulation. The inhibitory role of Gallic acid towards mast cell derived inflammatory allergic reactions can be attributed to the fact that it prevents the release of histamine, heparin and also blocks the cytokine expression which are actually involved in cell signaling (Kim et al., 2006).

GA also plays an important role in maintaining the skin health. GA exhibits anti-melanogenic properties. It acts as a skin whitening agent by controlling the process of melanogenesis. The ultra-violet induced skin-pigmentation process is controlled by Gallic acid that serves as a melanogenesis inhibitory agent. The enzyme tyrosinase is responsible for this biological process and its activity is subdued by Gallic acid (KIM, 2007). Due to these properties, GA is used as an additive in the cosmetic formulations as an anti-pigmentation agent.

GA can also display excellent antimicrobial activity against a wide range of pathogens such as *Pseudomonas* strains, *Escherichia coli* strains, *Staphylococcus aureus*, *Klebsiella pneumonia*, etc. by producing some irreversible changes in the membrane properties of the host cell. It not only acts as a potential antimicrobial agent but also enhances the efficacy of other antimicrobial drugs by inhibiting drug efflux pumps which are the major cause of development of antimicrobial resistance to any particular drug and thereby alleviate the efficacy of the drug (Borges et al., 2013).

Gallic acid and its n-alkyl esters are known to exhibit promising anti-viral properties too. These compounds display virucidal properties against a number of viruses viz. Herpes Simplex Virus type 1 (HSV-1), Human Immunodeficiency Virus (HIV-1), Human Rhinoviruses (HRVs) etc. which are responsible for causing mild to severe viral infections in humans. The replications of HRVs which are responsible for common cold are inhibited by GA through virus absorption (Choi et al., 2010). GA and pentylgallate is found to be effective against HSV-1 virus (Kratz et al., 2008).

GA also has an important industrial application as a food additive in various food stuffs like baked items, processed and packaged food items, candy, chewing gums etc. Due to its antioxidant property, it protects fats and oils from rancidity and spoilage. Propyl gallate, lauryl gallate, octylgallate, dodecyl gallate etc. are some of the esters of GA which are used as food preservatives. Industrial applications of GA in tanning of leather, paper making, in manufacturing of inks and paints are also very significant. Gallic acid can produce pure black color in reaction with Iron (II) sulphate due to the formation of iron gallate. So it is used as the chief ingredient in Iron Gall ink (Ramamurthy et al., 2014 & Akhila et al. 2022).

Structural Aspects of Gallic Acid

The physical and chemical properties of a material largely depend upon the internal arrangement of the constituting atoms/molecules in the solid state. Therefore, knowledge of the three-dimensional structure of a material is a prerequisite for structure-property correlations and for designing new functional materials. This is because of the fact that the solid crystalline form of the material dictates the physical properties, such as solubility, density, melting point, boiling point, etc., and consequently can greatly influence the manufacturing processes, bioavailability, thermal/chemical stability as well as the performance of drug products.

With the advent of modern methods of characterization and wider availability of powerful X-ray diffraction instruments, the thrust for development of molecular crystalline solids with specific properties have increased tremendously. Gallic acid, chemically 3,4,5-trihydroxybenzoic acid is a multi-functional organic molecule consisting of carboxylic acid (–COOH) and hydroxyl (–OH) moieties, and is known to exist in several polymorphic modifications. It is found that the biological activities of gallic acid are dependent on its structural characteristics. In this regard, structural elucidation of GA appears to be an important criterion for better understanding of its biological functions.

Several crystalline forms of gallic acid, including anhydrous or solvate/hydrate formation and their polymorphic forms have been reported recently (Tables 3, 4 & 5). Hirun et al. have reported the crystal structure of anhydrous gallic acid for the first time. Single crystal X-ray diffraction studies have shown that gallic acid has a planar structure and all the three hydroxyl H-atoms on the phenyl ring are oriented in the same direction. Both intra- and intermolecular hydrogen bonds are observed in the crystal structure of GA. There occurs two intramolecular O-H...O hydrogen bonds between the adjacent –OH moieties in the ring. In addition, several intermolecular hydrogen bonding forces are also operative resulting in a supramolecular architecture. A few more crystal structures of anhydrous gallic acid crystallized in different space groups (or polymorphs, *vide infra*) have also been reported. Table 3 collects the crystallographic data of anhydrous gallic acids available in Cambridge Crystallographic Database (CSD).

Table 3. Crystal data of anhydrous gallic acid molecules, $C_7H_6O_5$

Compound: Gallic acid (3,4,5-trihydroxy benzoic acid)							
Crystal System	Mono-clinic	Mono-clinic	Mono-clinic	Mono-clinic	Mono-clinic	Mono-clinic	Tri-clinic
Space Gr.	C2/c	C2/c	C2/c	C2/c	P21/c	C2/c	P-1
T (K)	90	293	298	100	298	293	298
a(Å)	25.690 (4)	25.629 (2)	25.650 (0)	25.675 (4)	5.230 (0)	25.685 (3)	7.319 (0)
b(Å)	4.895 (0)	4.921 (0)	4.922 (0)	4.905 (0)	5.265 (0)	4.927 (0)	8.254 (0)
c(Å)	11.097 (2)	11.222 (0)	11.230 (0)	11.115 (2)	24.793 (0)	11.251 (1)	11.715 (0)
α (°)	90.00	90.00	90.00	90.00	90.00	90.00	100.47 (0)
β (°)	105.75 (0)	106.25 (0)	106.28 (0)	105.77 (0)	102.11 (0)	106.23 (0)	90.23 (0)
γ (°)	90.00	90.00	90.00	90.00	90.00	90.00	90.98 (0)
R-Factor (%)	3.40	3.18	3.73	4.00	2.59	3.56	4.82
Reference	Zhao et al., 2011	Hirun et al., 2011	Clarke et al., 2013	Ermer et al., 2012.	Braun et al., 2013	Braun et al., 2013	Braun et al., 2013

(Source: CSD)

Polymorphs, Solvates/Hydrates, Organic Salts and Co-Crystals

The controlled preparation and characterization of different crystals from the same multicomponent substance is one of the major issues of modern crystal engineering and solid state chemistry. These materials may take up various solid forms depending upon the nature and the stoichiometry of the chemical components present in the solid. Usually, multicomponent crystalline solids are categorized as salts, solvates, polymorphs, hydrates or co-crystals (Figure 2).

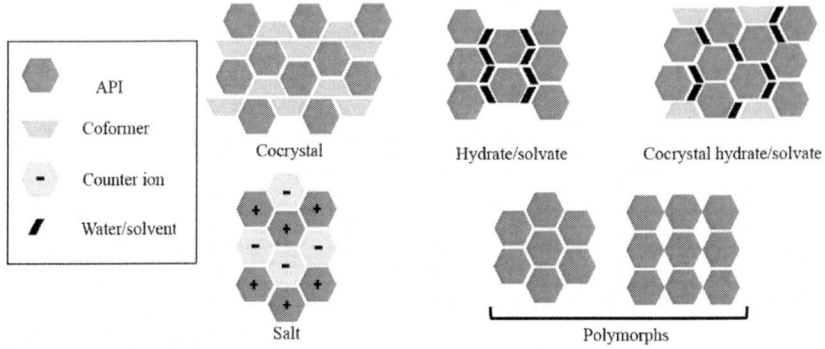

Figure 2. Classification of multicomponent crystalline solids.

Polymorphism is the existence of the same chemical substance in more than one crystalline modification for molecular chemistry and is regarded as a subset of supramolecular isomerism i.e., the same molecular components generate different supramolecular synthons. In other words, materials with the same chemical composition differ in the lattice structures and/or molecular conformation. Polymorphism is a complex phenomenon and its occurrence is more frequent in compounds that are conformationally flexible and also contain groups that are able to form strong hydrogen bonds such as –OH, –NH$_2$, –COOH and –CONH$_2$. *Pseudo-polymorphism* is a related term that has been coined to categorize solvates, especially in the context of pharmaceutical solids. It refers to crystalline forms with solvent molecules as an integral part of the structure. The pseudo-polymorphs are obtained in crystalline forms that differ in nature or stoichiometry of included solvent molecules. Pseudo-polymorph is a commonly used term by researchers but is ambiguous and causes controversy in literature. The term *solvate* is thus more convenient to describe different solvent containing crystalline forms whilst a *hydrate* contains water as the solvent of crystallization within the crystal structure framework. These are then also distinct in terms of physical properties from the pure anhydrous material.

The design and synthesis of co-crystals have received great interest among the researchers of pharmaceutical and chemical sciences as well as from drug regulatory agencies in recent years. The formation of these homogeneous crystalline solids is primarily based on the concept of supramolecular chemistry. Co-crystallization of Active Pharmaceutical Ingredient (APIs) or drug materials offers a great opportunity for developing new drugs with enhanced physical and pharmacological properties such as solubility, stability, bioavailability etc., during their therapeutic activity. Co-crystals provide multiple opportunities to modify the chemical/physical properties of an API without making or breaking its covalent bonds. Material Science research is constantly focusing on drug molecules and their most suitable forms for development in terms of stability and solubility. Most frequently employed chemical approaches for the purpose is the co-crystal formation. The constituent molecules of these co-crystals are connected with each other through various non-covalent/supramolecular interactions. Depending on the nature of interactions, the co-crystal former (a crystallizing agent), associated with the target molecule can be exchanged without compromising the covalent linkages present in the target molecule. This alters the composition of co-crystals and hence, the bulk properties of the crystalline solids. As such, the formation of co-crystals may turn out to be a new and efficient route to improve the physico-chemical properties of crystalline drug molecules.

Co-crystals are a class of homogenous crystalline systems in which stoichiometric amounts of two or more neutral molecules are held together under ambient conditions utilizing intermolecular forces of attraction. It involves the formation of supramolecular homo (I-II)/hetero (III-IV)-synthons (Figure 3) through various weaker non-covalent interactions like hydrogen bonding, π–π interactions or van der Waals forces. Thus, it is an achievable and effective method for improving the bulk properties of the targeted compound.

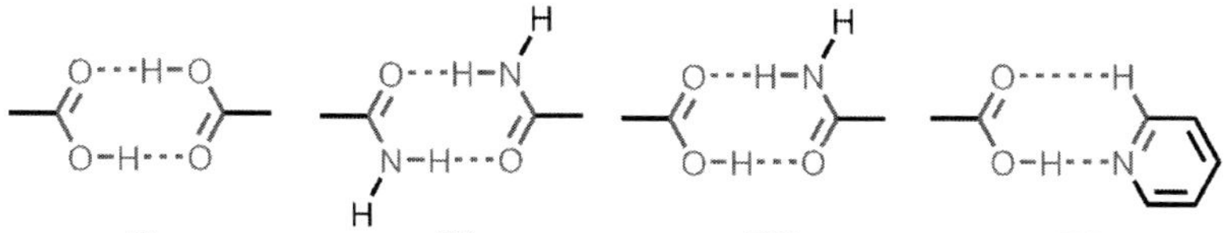

Figure 3. Supramolecular homo and hetersynthons of co-crystals.

Another similar term, 'organic salt' is interrelated to 'co-crystal' through a proton transfer pathway (Aaköry et al., 2007; Sedghiniya et al., 2019; Seliger et al., 2014). In fact, it is the location of the acidic proton (between the acid and the base) that determines the nature of the multicomponent crystal. Relocation of protons towards the base indicates greater probability of salt formation rather than co-crystal. The proton donating ability of an aromatic acid is largely influenced by the presence of electron donating/withdrawing substituents. It has been found that electron-withdrawing groups increase the acidity of aromatic acid whereas electron-donating groups reduce the acidity. Thus, the tendency of a hydrogen bond accepting molecule for co-crystal or salt formation depends upon the nature and location of substituents present in the acid molecule. Mohamed et al. (2009) have studied the relative influence of pyridine–carboxylic acid and pyridinium–carboxylate synthons on crystal packing of the binary solids.

Table 4. Crystal data of gallic acid monohydrates, $C_7H_6O_5 \cdot H_2O$

Compound: Gallic acid monohydrate (3,4,5-trihydroxy benzoic acid monohydrate)								
Crystal System	Mono-clinic	Mono-clinic	Mono-clinic	Mono-clinic	Mono-clinic	Mono-clinic	Mono-clinic	Tri-clinic
Space Gr.	P21/c	P2/n	P2/n	P21/c	P21/c	P21/c	P2/n	P-1
T (K)	295	296	293	296	123	123	250	150
a(Å)	5.794 (4)	14.150 (10)	14.158 (3)	9.794 (0)	7.607 (0)	9.757 (0)	14.172 (3)	7.121 (0)
b(Å)	4.719 (5),	3.622 (9)	3.624 (1),	3.612 (0)	3.641 (0)	3.558 (0)	3.606 (0)	7.509 (0)
c(Å)	28.688 (5)	15.028 (10)	15.036 (3)	21.591 (1)	26.792 (0)	21.517 (0)	15.022 (4)	15.588 (1)
α (°)	90.00	90.00	90.00	90.00	90.00	90.00	90.00	100.58 (0)
β (°)	95.08 (3)	97.52 (7)	97.60 (3)	91.27 (0)	98.42 (0)	91.34 (0)	97.58 (0)	91.11 (0)
γ (°)	90.00	90.00	90.00	90.00	90.00	90.00	90.00	111.45 (0)
R-Factor (%)	4.20	4.92	5.34	3.91	4.18	3.50	3.91	4.13
Reference	Jiang et al., 2000.	Okabe et al., 2001	Billeset al., 2007	Demirtas et al., 2011	Braun et al., 2013	Hoseret al., 2017.	Ponce et al., 2018	Ponce et al., 2018

(Source: CSD)

Table 5. Crystal data of gallic acid solvates, $C_7H_6O_5$ solvent

Compound: Gallic acid solvate (3,4,5-trihydroxy benzoic acid solvate)							
Crystal System	GA:(methanol): (hydrate)	GA:(acetamide)	GA:(acetamide)	GA:(acetamide)	GA:(acetamide)	GA:3(acetamide)	GA: (pyridine)
	Monoclinic	Monoclinic	Monoclinic	Monoclinic	Monoclinic	Triclinic	Monoclinic Triclinic
Space Gr.	P21/n	P21/c	P21/n	P21/n	C2/c	P-1	P21/n
T (K)	100	110	110	110	120	110	293
a(Å)	3.654 (0)	9.640 (0)	4.263 (0)	5.191 (0)	28.342 (6)	7.186 (0)	9.335 (1)
b(Å)	27.246 (1)	10.448 (0)	22.500 (1)	18.999 (0	4.903 (0)	7.931 (0)	10.435 (2)
c(Å)	9.243 (0)	10.403 (0)	10.049 (0)	10.073 (0)	20.113 (4)	14.601 (0)	11.858 (1)
α (°)	90.00	90.0	90.0	90.0	90.00	82.64(0)	90.00
β (°)	94.49 (0)	107.91 (0)	92.87 (0)	100.33 (0)	134.53 (4)	82.39 (0)	107.63 (0)
γ (°)	90.00	90.0	90.0	90.0	90.00	87.09(0)	90.00
R-Factor (%)	5.55	2.97	3.64	3.54	3.38	3.87	6.58
Reference	Thomas et al., 2013.	Kaur et al., 2012.	Kaur et al., 2012.	Kaur et al., 2012.	Kaur et al., 2012.	Kaur et al., 2012.	Dong et al., 2011.

(Source: CSD)

Gallic Acid Co-Crystals in Pharmaceutical Research

Co-crystals have gained considerable interest in pharmaceutical research due to their ability to improve physicochemical characteristics of an API. Due to the presence of potential hydrogen bond donor and acceptor sites, GA appears to be an excellent candidate for co-crystal formation. This biologically relevant molecule possesses various pharmacological activities, such as antioxidant (Souza et al., 2011), antibacterial (Chanwitheesuk et al., 2007), antitumor (Liu et al., 2011), antiviral (Ozcelik et al., 2011) properties etc. Besides these, GA has no toxicity and side effects even in large doses. Consequently, GA has received much attention in the field of pharmacological research (Badhani et al., 2015).

On the other hand, low thermal stability, large particle size and poor solubility during absorption, are some of the issues due to which pharmacological activities of GA are significantly diminished. The therapeutic effectiveness of these materials greatly depends on their solubility, because poor solubility can cause low bioavailability of the same (Daneshfar et al., 2008; Lu & Lu 2007). Reports reveal that the solubility and dissolution rate of GA can be greatly enhanced by constructing GA-based co-crystals with the help of several co-crystal formers (CCFs) (Surov et al.,2018; Kaur et al., 2014; Kaur et al., 2016; Dabir et al., 2018). In recent times, an increasingly larger section of chemists has started looking into the structures of binary crystalline solids with a view of examining the enhanced pharmacological properties in them.

Na Xue et al. (2020), have successfully isolated two co-crystals of gallic acid with glutaric acid and succinimide (Figure 4) and studied their solubility, dissolution rate and oral bioavailability. In their study, they found that both these crystalline entities exhibited rapid dissolution as compared to that of neat GA. In particular, the solubility of the GA-succinimide system increased by more than 1.5 times. Further, *in vitro* α-glucosidase inhibition study reveals that both GA and glutaric acid demonstrated the inhibitory activity, while succinimide showed no such activity. However, the co-crystal of GA and glutaric acid presented a higher inhibitory effect than those of the two parent components. This enhanced inhibitory activity may be attributed to the synergism of the two components. Molecular docking simulation indicated that the GA-glutaric acid complex deeply entered the active cavity of the α-glucosidase in the form of a supramolecule, which made the guest-enzyme binding configuration more stable.

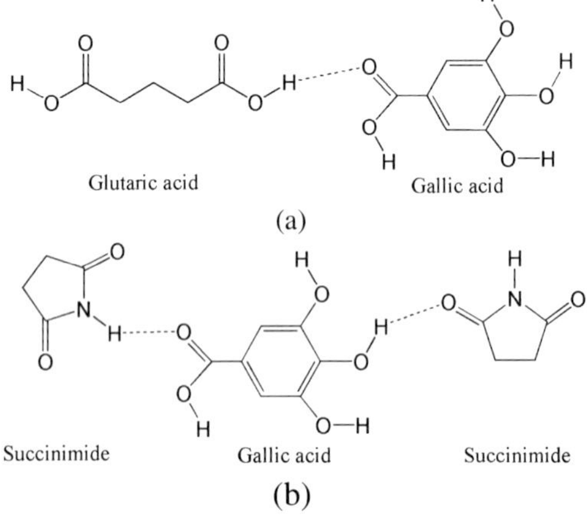

Figure 4. Co-crystals of (a) Gallic acid and glutaric acid (1:1) and (b) Gallic acid and succinimide (1:2).

Co-crystals of biologically active 1-[5-(3-chloro-phenylamino)-1,2,4-thiadiazol-3-yl]-propan-2-ol (TDZ) and GA (in different stoichiometric ratios) are also found to exhibit improved solubility behavior (Surov et al., 2018). The co-crystallized species *viz.*, [2(TDZ):GA], [TDZ:GA] and [TDZ:GA:H$_2$O] were isolated from the acetonitrile/water mixture (Figure 5). The dissolution study shows that the hydrated species, [TDZ:GA:H$_2$O] (1:1:1) was about 6.6 times more soluble than the parent TDZ at pH 2.0 and 25°C.

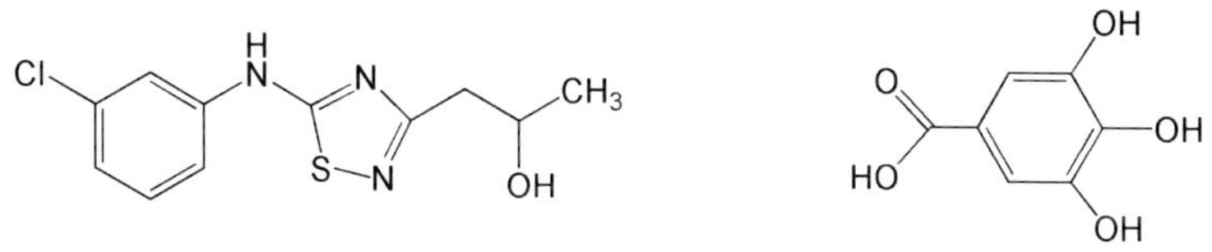

1-[5-(3-chloro-phenylamino)-1,2,4-thiadiazol-3-yl]-propan-2-ol   Gallic acid

Figure 5. (1:1) Co-crystal of GA and TDZ.

Flucytosine or 5-fluorocytsine (5-FC) is a well-known drug and is used to treat fungal infections in the body. This drug is moisture sensitive and thus creates issues during its storage. To mitigate this issue, 5-FC has been co-crystallized with GA and few more organic acids and the stabilities of the resulting co-crystals/salts were evaluated (Nechipadappu et al. 2018) (Figure 6). It was found that 5-FC-GA co-crystal did not experience any hydration under accelerated humidity conditions (both 70-75% RH and 90-95%RH) at ambient temperature (~30°C). Furthermore, the synthesized co-crystal is found to be stable for 2 months at ambient conditions (~30°C, 60-65% RH).

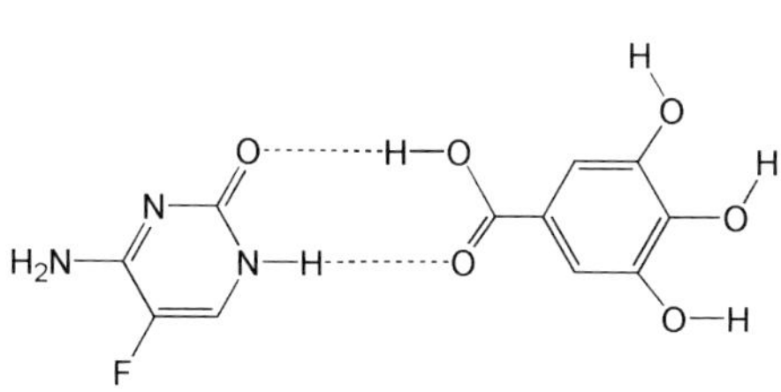

5-fluorocytosine   Gallic acid

Figure 6. Supramolecular hetero synthons observed in (1:1) co-crystal of GA and 5-FC.

Isoniazid, also known as isonicotinic acid hydrazide (INH), is used with other medications to treat tuberculosis (TB) infections. In tuberculosis and other diseases, tissue inflammation and free radicals burst from macrophages resulting in oxidative stress. These free radicals cause pulmonary inflammation if not countered by antioxidants. In this perspective, Mashhadi et al. (2014), have successfully isolated four crystalline adducts of isoniazid by co-crystallizing it with gallic acid, 2,3-dihydroxybenzoic acid, 3,5-dihydroxybenzoic acid and 3-hydroxybenzoic acid in the stoichiometric ratio 1:1. The reason for choosing these hydroxy benzoic acids for co-crytsal formulation is their potential as antioxidants.

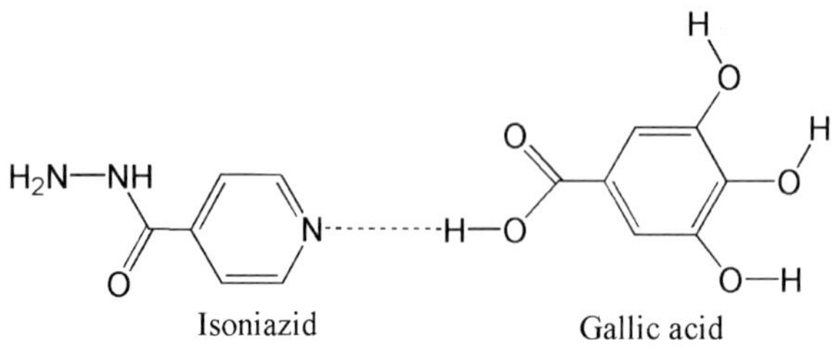

Figure 7. GA-INH (1:1) co-crystal.

Jia-Xi Song et al. (2015), have reported the preparation and structures of two lenalidomide co-crystals with gallic acid in the molar ratio 1:1. After the formation of co-crystals, their dissolution rate and apparent solubility were evaluated and were found to be enhanced. The high solubility of co-crystals can keep for 48 h without dropping. The result of phase solubility study indicates that gallic acid (GA) forms a 1:1 complex with lenalidomide in an aqueous solution, with a stability constant of 1713.2 L.mol^{-1}. Extensive hydrogen bonding interactions between gallic acid and lenalidomide are attributed to the formation of a binary adduct in the aqueous solution and subsequently we find the constant high solubility of the resulting product.

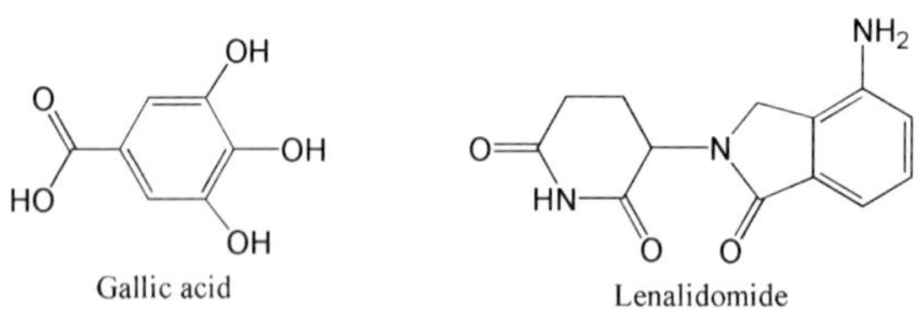

Figure 8. (1:1) Co-crystals of gallic acid and lenalidomide.

Metronidazole (MNZ) is a widely used water-soluble antimicrobial drug that can cause increased plasma concentration after its administration. The high plasma concentration in the body may lead to serious encephalopathy.

Pharmacognosy of Gallic Acid and Its Co-Crystals 167

Zhang et al., have tried to optimize these properties of MNZ by adopting the co-crystallization strategy. Thus, they have prepared a co-crystal of MNZ with GA in the molar ratio 1:1 and performed a comparative study on the dissolution and absorption behaviors of neat MNZ and MNZ–GA co-crystal. The results indicate that the MNZ–GA co-crystal exhibits lower C_{max} and longer T_{max} than pure MNZ, suggesting that the prepared MNZ–GA co-crystal may find application in the effective formulation of MNZ.

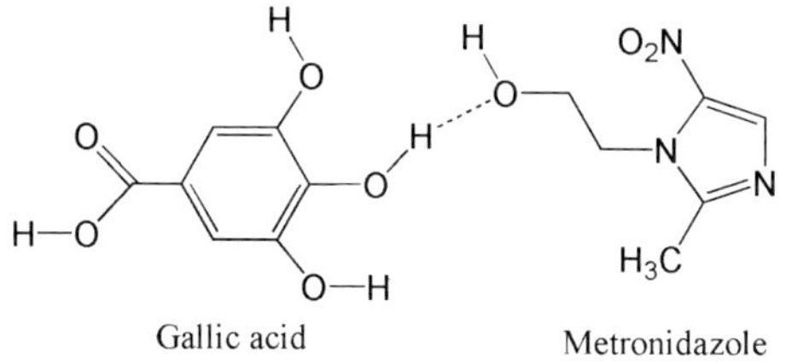

Figure 9. (1:1) Co-crystal of GA-MNZ.

Goswami et al., (2021) have successfully isolated a (1:1) co-crystal involving GA and 4-CNpy molecule. Crystal structure analysis reveals that GA and 4-CNpy molecules are connected with each other through various non-covalent interactions to form a 3-D supramolecular assembly. Antioxidant experiments indicated that the binary co-crystal, GA-CNpy could enhance its antioxidant property compared to that acquired by GA and 4-CNpy alone. The *in vitro* antimicrobial activity of the co-crystal was also evaluated, and was found highly active against several human pathogens including *Escherichia coli*. The inhibitory role of GA-CN pycocrystal against drug efflux pump AcrB in *E. coli* was further established by molecular docking analysis.

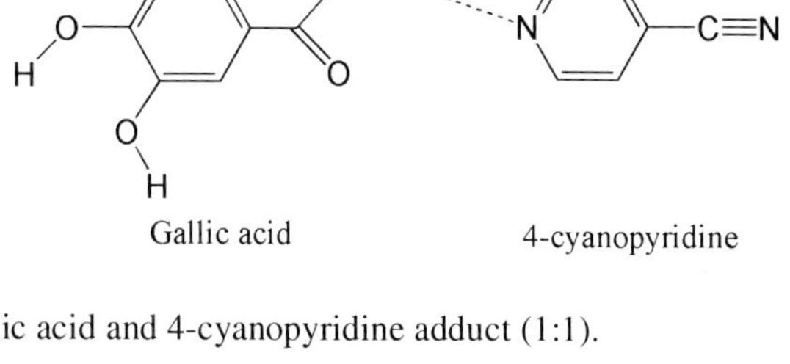

Figure 10. Gallic acid and 4-cyanopyridine adduct (1:1).

Conclusion

The above review of literature establishes gallic acid and its co-crystals as a potential molecule for further study in its application as a health promoting molecule. Furthermore, it has been observed that this compound has already been in use in ayurvedic formulations for a long time. This shows that it has no potential health hazard. The pharmacological application of gallic acid is diverse and is thus considered as a suitable candidate for further analysis of its pharmacological properties.

References

Aakeröy CB, Fasulo ME, Desper J. *Mol. Pharmaceutics*, (2007) 4:317.
Aher AN, Pal SC, Yadav SK, Patil UK. *Asian J Chem* (2010) 22: 3429.
Akhila K, Ramakanth D, Gaikwad KK. *Journal of Coatings Technology and Research* (2022) 19:1493–1506.
Badhani B, Sharma N, Kakkar R, *RSC Adv.* (2015) 5:27540.
Baiano A. In *Handbook on Gallic Acid*, edited by Thompson MA and Collins, PB. (2013) 177. Nova Science Publishers, Inc.
Bajpai B, Patil S. *Braz J Microbiol* (2008) 39(4): 708.
Basu A, Penugonda K. *Nutrition Reviews*. (2009) 67(1):49.
Billes F, Mohammed-Ziegler I, Bombicz P. *Vibrational Spectroscopy* (2007) 43:193.
Borde VU, Pangrikar PP, Tekale SU. *Recent Research in Science and Technology*. (2011) 3(7): 51.
Borges A, Ferreira C, Saavedra MJ, Simoës M. *Microbial Drug Resistance* (2013). Mary Ann Liebert, Inc.,
Borges A, Saavedra MJ, Simões M. *Biofouling*. (2012) 28(7):755.
Braun DE, Bhardwaj RM, Florence AJ., Tocher DA, Price SL. *Cryst. Growth Des.* (2013) 13:19.
Chanwitheesuk A, Teerawutgulrag A, Kilburn JD, Rakariyatham N. *Food Chem.* (2007) 100:1044.
Chen HM, Wu YC, Chia YC, Chang, FR, Hsu HK, Hsieh YC, Chen CC, Yuan SS. *Cancer Lett.*(2009) 286:161.
Choi HJ, Song JH, Bhatt LR, Baek SH. *Phytother. Res.* (2010) 24:1292–1296.
Clarke HD, Arora KK, Wojtas L, Zaworotko MJ. *Cryst. Growth Des.* (2011) 11:964.
Dabir TO, Gaikar VG, Jayaraman S, Mukherjee S. *Fluid Phase Equilib.*(2018) 456:65.
Dalla PC, Padovani G, Mainente F, Zoccatelli G, Bissoli G, Mosconi S, Veneri G, Peruffo A, Andrighetto G, Rizzi C, Chignola, R. *Cancer Lett.* (2005) 226:17.
Daneshfar A, Ghaziaskar HS, Homayoun N. *J. Chem. Eng. Data* (2008) 53:776.
de la Rosa LA, Alvarez-Parrilla E, Gonzalez-Aguilar GA. *Fruit and Vegetable Phytochemicals- Chemistry, Nutritional Value, and Stability*, (2010) 1st Ed, Wiley-Blackwell: Ames, IA, USA.

Demirtas G, Dege N, Buyukgungor O. *Acta Crystallogr., Sect.E* (2011) 67:o1509.
Deschamps AM, Lebeault J.M. *Biotechnol. Lett.* (1984) 6:237–242.
Dong F-Y, Wu J, Tian H-Y, Ye Q-M, Jiang R-W. *Acta Crystallogr., Sect. E* (2011) 67:o3096.
Ermer O, Neudorfl J. *Helv. Chim. Acta,* (2012) 95:872.
Faried A, Kurnia D, Faried LS, Usman N, Miyazaki T, Kato H, Kuwano H. *International Journal of Oncology* (2007) 30:605.
Goswami S, Ghosh A, Borah K, Mahanta A, Guha AK, Bora SJ. *J. Mol. Struct.* (2021) 1225:129279.
Halliwell B, Aeschbach R, Löliger J, Aruoma OI. *Food Chem. Toxicol.* (1995) 33:601.
Hill AF. *Economic Botany: A Textbook of Useful Plants and Plant Products*(1952) (2nd Ed.) McGraw-Hill Book Company Inc, NY.
Hirun N, Saithong S, Pakawatchai C, Tantishaiyakul V. *Acta Crystallogr., Sect. E* (2011) 67:o787.
Hoser AA, Sovago I, Lanza A, Madsen A. *Chem. Commun.* (2017) 53:925.
How PS, Ellis JA, Spencer JPE, Williams C. *Appetite* (2008) 51(3):754.
Jiang R-W, Ming D-S, But PPH, Mak TCW. *Acta Crystallogr., Sect.C* (2000) 56:594.
Kahkeshani N, Farzaei F, Fotouhi M, Alavi SS, Bahramsoltani R, Naseri R, Momtaz S, Abbasabadi Z, Rahimi R, Farzaei MH, Bishayee A. *Iran J Basic Med Sci.* (2019) 22(3):225-237.
Karamaæ MA, Kosiñska PRB. *Pol J. Food Nutr Sci.* (2005) 14:165.
Kaur M, Velmurugan B, Rajamanickam S, Agarwal R, Agarwal C. *Pharm. Res* .(2009) 26:2133.
Kaur R, Cherukuvada S, Managutti PB, Row TNG. *Cryst Eng Comm* (2016) 18:3191.
Kaur R, Perumal SSRR, Bhattacharyya AJ, Yashonath S, Row TNG, *Cryst. Growth Des.* (2014) 14:423.
Kaur R, Row TNG. *Cryst. Growth Des.* (2012) 12:2744.
Kaur S, Husheem M, Arora S, Pirkko LH, Kumar S. *J. Ethnopharmacol* (2005) 97(1):15.
Khadem S, Marles RJ. *Molecules* (2010) 15(11):7985.
Khan AN, Hassan MN, Khan RH. *J. Mol. Liq.* (2019) 285:27–37.
Khanbabaee K, Ree VT. *Nat. Prod. Rep.* (2001) 18:641.
Kim S-H, Jun C-D, Suk K, Choi B-J, Lim H, Park S, Lee SH, Shin H-Y, Kim D-K, Shinet T-Y. *Toxicological Sciences* (2006) 91(1):123.
KIM Y-J. *Biol. Pharm. Bull.* (2007) 30(6):1052.
Korani MS, Farbood Y, Sarkaki A, Moghaddam HF, Mansouri MT. *European Journal of Pharmacology* (2014) 733:62.
Kratz JM, Andrighetti-Fröhner CR, Kolling DJ, Leal PC, Cirne-Santos CC, Yunes RA, Nunes RJ, Trybala E, Bergström T, CPP Frugulhetti I, Monte Barardi CR, Oliveira Simões CM. *Mem Inst Oswaldo Cruz*, Rio de Janeiro (2008) 103(5): 437.
Lafay S, Gil-Izquierdo A. *Phytochem. Rev.* (2008) 7:301.
Liang J, Huang X, Ma G.*RSC Adv.* (2022) 12(45):29197.
Liu KC, Huang AC, Wu PP, Lin HY, Chueh FS, Yang JS, Lu CC, Chiang JH, Meng M, Chung J G. *Oncol. Rep.* (2011) 26:177.
Liu ZJ, Schwimer J, Liu D, Lewis J, Greenway FL, York DA, Woltering EA. *Phytother. Res.* (2006) 20:806.

Lu L-L, Lu X-Y. *J. Chem. Eng. Data* (2007) 52:37.
Lu Z, Nie G, Belton PS, Tang H, Zhao B. *Neurochemistry International* (2006) 48:263.
Manach C, Scalbert A, Morand C, Rémésy C. and Jimenez L. *Am J Clin Nutr* (2004)79:727–47.
Mashhadi SMA, Yunus U, Bhatti MH, Tahir MN. *J. Mol. Struct.* (2014) 1076:446.
Mohamed S, Tocher DA, Vickers M, Karamertzanis PG, Price SL. *Cryst. Growth Des.* (2009) 9:2881.
Nandi A, Yan L-J, Jana CK, Das N. *Hindawi Oxidative Medicine and Cellular Longevity Volume* (2019).
Nayeem N, Asdaq SMB, Salem H, Said AA. *J App Pharm.* (2016) 8(2):1000213.
Nechipadappu SK, Ramachandran J, Shivalingegowda N, Lokanath NK, Trivedi DR, *New J. Chem.* (2018) 42:5433.
Okabe N, Kyoyama H, Suzuki M. *Acta Crystallogr., Sect.E* (2001) 57:o764.
Olmedo-Juárez A, Briones-Robles TI, Zaragoza-Bastida A, Zamilpa A, Ojeda-Ramírez D, Mendoza de Gives P, Olivares-Pérez J, Rivero-Perez N. *Microb Pathog.* (2019) 136:103660.
Orhan IE. *Bioimpacts* (2014) 4(3):109.
Ozcelik B, Kartal M, Orhan I, *Pharm. Biol.* (2011) 49:396.
Pandey KB, Rizvi SI. *Oxidative Medicine and Cellular Longevity* (2009) 2:270.
Ponce A, Zavalij Peter Y, Eichhorn BW. *CSD Communication (Private Communication)* (2018).
Pourrat H, Regerat F, Pourrat A, Jean D. *J. Ferment. Biotechnol.* (1985) 63:401.
Ramamurthy G, Krishnamoorthy G, Sastry TP, Mandal AB. *Clean Technol. Environ. Policy* (2014) 16:647.
Saavedra MJ, Borges A, Dias C, Aires A, Bennett RN, Rosa ES, Simões M. *Med Chem.* (2010) 6(3):174.
Sandhya T, Lathika KM, Pandey BN, Mishra KP. *Cancer Lett.*(2006) 231(2):206.
Sedghiniya S, Soleimannejad J, Janczak J. *Acta Crystallogr., Sect. C* (2019) C75:412.
Seliger J, Žagar V. *Phys. Chem. Chem. Phys.* (2014) 16:18141.
Seth M, Chand S. *Process Biochem.* (2000) 36(1–2):39.
Shabani S, Rabiei Z, Amini-Khoei H. Exploring the multifaceted neuroprotective actions of gallicacid: A review. *Int. J Food Prop.* (2020) 23(1):736–752.
Shahriar K, Robin JM. *Molecules* (2010) 15:7985.
Singh MP, Gupta A, Sisodia SS. *Int. J. of Pharmacognosy and Phytochem. Res.* (2018) 10(4):132.
Sneha C, Lesley R, Lesley RV, Vinod K, Vikas B. *Pharm Pat Anal.* (2015)4(4):305.
Song J-X, Chen J-M, Lu T-B. *Cryst. Growth Des.* (2015) 15(10):4869.
Soong YY, Barlow PJ. (2004) *Food* 88:411.
Souza BWS, Cerqueira MA, Martins JT, Quintas MAC, Ferreira ACS, Teixeira JA, Vicente AA, *J. Agric. Food Chem.* (2011) 59:5589.
Surov AO, Churakov AV, Proshin AN, Dai XL, Lu T, Perlovich GL. *Phys. Chem. Chem. Phys.*(2018) 20:14469.
Thomas SP, Kaur R, Sankolli R, Kaur J, Nayak SK, Row TNG. *J. Mol. Struct.* (2013) 30:88.
Thompson MA, Collins PB. *Handbook on Gallic Acid: Natural Occurrences, Antioxidant Properties and Health Implications* (2013) Nova Science Publishers, Inc.

Tougu V, Tiiman A, Palumaa P. *Metallomics* (2011) 3:250.
Tyler VE, Brady LR, Robbers JE, (1988) *Pharmacognosy.* Lee and Febiger.
Verma S, Singh A, Mishra A. Gallic acid: Molecular rival of cancer, *Environmental Toxicology and Pharmacology* (2013) 35(3):473.
Xu Y, Tang G, Zhang C, Wang N, Feng Y. *Molecules* (2021) 26:7115.
Xue N, Jia Y, Li C, He B, Yang C, Wang J. *Molecules* (2020) 25:1163.
You BR, Moon HJ, Han YH, Park WH. *Food and Chemical Toxicol.* (2010) 48(5):1334.
Youdim KA, Joseph JA. *Free Radical Biology and Medicine.* (2001) 30:583.
Zhao J, Khan IA, Fronczek FR. *Acta Crystallogr., Sect. E* (2011) 67:o316.
Zheng K, Li A, Wu W, Qian S, Liu B, Pang Q. *J. Mol. Struct.* (2019) 1197:727.

Biographical Sketches

Sanchay Jyoti Bora, MSc, PhD

Affiliation: Pandu College, Guwahati 781012, Assam, India

Research and Professional Experience:

- Nineteen years of research experience. Areas of research include Supramolecular Chemistry, X-ray Crystallography, Organic Co-crystals, MOFs, MOF-based nanocomposites, Catalysis, etc.
- Seventeen years of teaching experience (Inorganic Chemistry & Analytical Chemistry).

Professional Appointments: Associate Professor and HoD, Chemistry, Pandu College

Honors:

1. Focus Area Science and Technology (FAST) Summer Fellowship (2019) by Indian Science Academies: Research Supervisor: Prof. G. Ranga Rao, Department of Chemistry, IIT Madras, Chennai, India.
2. Summer Research Fellowship (SRF) (2014) by Indian Science Academies: Research Supervisor: Dr. S.N. Achary, Bhaba Atomic Research Centre (BARC), Mumbai, India.
3. AsCA '10 Travel Support Prize: For presenting a paper on the 10th Conference of the Asian Crystallographic Association (AsCA '10) at Busan, South Korea.

4. AsCA'07 Travel Support Prize: For presenting a paper on the 8th Conference of the Asian Crystallographic Association (AsCA '07) at Taipei, Taiwan.

Publications from the Last 3 Years:

Journals:

1. Structure-Property Correlation in Gallic Acid and 4-Cyanopyridine Cocrystal and Binding Studies with Drug Efflux Pump in Bacteria, *J. Mol. Struct.* 2021,1225, pp. 129279.
2. Trinuclear Mn 2+ /Zn 2+ based microporous coordinationpolymers as efficientcatalysts for ipsohydroxylation of boronic acids. *Dalton Trans.* 2020, 49, pp. 5454.
3. Ditopic carboxylate containing zigzag chain polymers with tetrahedral Co(II) and Zn(II) nodes. *J. Mol. Struct.* 2020, 1217, pp. 128434.
4. Isostructurality of complexes of the type tetraaquabis(isonicotinato)metal(II), *J. Indian Chem. Soc*. 2019, 96, pp. 317.

Text Books:

1. *Basic Analytical Chemistry*, for the UG 3rd Semester CBCS Honours Course, 2020; Sampriti Publication, Guwahati, Assam, ISBN: 978-81-946130-1-5 authored by Sanchay Jyoti Bora & Purabi Sarmah.

2. *Analytical Methods in Chemistry*, authored by Sanchay Jyoti Bora & Purabi Sarmah for UGCBCS final year students (both Honors and Regular), 2022, Vishal Publication & Co., Jalandhar-Delhi, India, ISBN: 978-93-91247-08-9.

Riju Kakati Sarma, M.Sc., Ph.D.

Affiliation: Pandu College, Guwahati 781012, Assam, India

Research and Professional Experience:

- More than 20 years of research experience in Gauhati University in aerobiology and antimicrobial activity of higher plant extracts.

- Twenty-six years of teaching experience in the UG level.

Professional Appointments:

- Associate Professor, Department of Botany, Pandu College, Guwahati.
- HoD Botany, Pandu College, Guwahati.

Publications from the Last 3 Years:

Journals:

1. Enumeration of aquatic mycoflora of Deepor Beel, a Ramsar site of Guwahati, Assam, India, *Research Journal of Social & Life Sciences.* 2019, pp. 103.

Text Books:

1. *Ecology and Taxonomy (Botany) in the Assamese Medium*, for the UG 2nd Semester CBCS, Generic Course, in June 2021. ISBN: 978-93-91158-10-1.

Purabi Sarmah, MSc, PhD

Affiliation: Nalbari College, Nalbari 781335, Assam, India

Research and Professional Experience:

- Eight years of research experience in Gauhati University as well as in the National Chemical Laboratory, Pune in the field of catalysis
- Thirteen years of teaching experience in UG course (Inorganic/ Spectroscopy/Analytical Chemistry).

Professional Appointments:

- Worked as an assistant professor in the Department of Chemistry, Bineswar Brahma Engineering College, Kokrajhar, BTAD, Assam from Aug. 2010 to Feb. 2015.

- Presently working as an assistant professor in the Department of Chemistry, Nalbari College, Nalbari, Assam since Feb., 2015.

Honors: Received Best Poster Award at CRSI (Chemical Research Society of India) Conference in 2007, held at IIT, Guwahati.

Publications from the Last 3 Years:

Text Books:

1. *Basic Analytical Chemistry*, for the UG 3rd Semester CBCS Honours Course, 2020; Sampriti Publication, Guwahati, Assam, ISBN: 978-81-946130-1-5 authored by Sanchay Jyoti Bora & Purabi Sarmah.
2. *Analytical Methods in Chemistry*, authored by Sanchay Jyoti Bora & Purabi Sarmah for UGCBCS final year students (both Honors and Regular), 2022, Vishal Publication & Co., Jalandhar-Delhi, India, ISBN: 978-93-91247-08-9.

Index

#

3,4,5-trihydroxybenzoic acid, vii, 31, 33, 53, 100, 158

A

aberrant cells, vii, 31, 71
accounting, viii, 45, 71, 78, 116
acetic acid, 13, 14, 15, 20, 134
acetone, 33, 134, 139
acetonitrile, 18, 24, 134, 165
acidic, 3, 13, 97, 98, 111, 139, 140, 162
adenocarcinoma, 48, 57, 69, 124, 126
adhesion, 82, 90, 101, 120, 125, 142
adverse effects, 47, 72, 156
alkaloids, 64, 137, 150
amino, 63, 64, 102, 140
amino acid, 63, 64, 102
angiogenesis, viii, 32, 113, 118, 122, 127, 157
antibacterial, vii, viii, xi, 32, 34, 50, 58, 78, 82, 83, 88, 89, 103, 111, 112, 124, 148, 154, 155, 163
antibiotic, 58, 82, 89, 124, 154
anti-cancer, vi, vii, viii, ix, x, xi, 1, 3, 23, 32, 35, 42, 43, 46, 47, 49, 50, 52, 53, 56, 58, 59, 60, 63, 67, 68, 69, 71, 72, 74, 75, 77, 78, 79, 80, 88, 102, 109, 111, 112, 115, 119, 120, 121, 123, 124, 125, 126, 129, 130, 131, 136, 141, 142, 145, 150, 153, 154, 155, 157
anti-diabetic, v, x, 44, 77, 78, 83, 93, 94, 95, 96, 100, 101, 104, 105, 107, 108, 111, 154, 155

anti-inflammatory, vii, viii, x, 1, 3, 32, 35, 42, 49, 63, 73, 77, 78, 84, 85, 86, 90, 91, 93, 100, 102, 103, 104, 111, 150, 154
antimicrobial, x, 23, 28, 59, 74, 77, 78, 79, 82, 88, 89, 102, 107, 146, 150, 153, 155, 157, 166, 167, 172
anti-myocardial, x, 77, 78
anti-obesity, viii, x, 32, 77, 78
antioxidants, vii, viii, x, xi, 1, 3, 22, 23, 24, 26, 28, 29, 32, 33, 34, 36, 42, 49, 50, 51, 56, 57, 59, 63, 67, 68, 69, 70, 71, 73, 74, 75, 77, 78, 79, 80, 81, 84, 86, 87, 88, 91, 93, 96, 98, 99, 100, 101, 103, 104, 105, 106, 107, 108, 111, 112, 113, 120, 123, 126, 130, 132, 134, 139, 143, 145, 146, 148, 149, 150, 153, 154, 155, 158, 163, 166, 167, 170
antitumor, xi, 79, 80, 86, 148, 154, 155, 163
antiviral, viii, xi, 32, 34, 44, 79, 89, 111, 124, 148, 154, 163
apoptosis, viii, xi, 32, 34, 40, 41, 46, 49, 50, 52, 53, 54, 55, 56, 58, 59, 60, 63, 68, 71, 79, 80, 86, 87, 91, 97, 98, 99, 100, 110, 112, 113, 114, 115, 116, 117, 118, 120, 121, 122, 124, 125, 126, 143, 157
apples, 98, 110, 151
arrest, viii, 32, 40, 54, 69, 80, 114, 116, 124, 125, 126
ascorbic acid, 21, 25, 81, 88, 124, 144
asthma, 108, 153, 157
astringent, 35, 81, 154, 155
atherosclerosis, 79, 86, 99
atoms, 18, 74, 139, 158, 159

Index

B

bacteria, 82, 88, 89, 100
base, ix, 5, 18, 57, 62, 63, 75, 102, 105, 111, 143, 146, 162
benefits, xi, 50, 70, 102, 103, 104, 111, 147, 149
benzene, 3, 64, 149
beverages, 63, 78, 103
bioavailability, viii, xii, 35, 36, 43, 46, 47, 57, 63, 70, 96, 100, 107, 111, 112, 130, 148, 158, 161, 164
biochemical processes, 32, 49, 67
biological activities, vii, x, 1, 77, 79, 84, 113, 159
biomolecules, 67, 68, 132, 155
black tea, 36, 90, 102, 106
blood, 33, 36, 37, 38, 54, 95, 101, 102, 103, 119, 122, 154
blood vessels, 37, 38, 122
body weight, 35, 36, 37, 84
bonding, 112, 140, 159, 161, 166
bonds, 7, 50, 159, 160, 161
brain, 38, 106, 143, 154, 156, 157
branching, 65, 66, 74
breakdown, 56, 64, 116
breast cancer, vii, viii, 31, 32, 37, 38, 39, 40, 43, 44, 62, 66, 78, 87, 88, 115, 124

C

cancer, v, vi, vii, viii, ix, x, xi, 1, 3, 31, 32, 34, 35, 37, 38, 39, 40, 41, 43, 44, 45, 46, 47, 48, 49, 50, 51, 52, 53, 54, 55, 56, 57, 58, 59, 60, 61, 62, 63, 66, 67, 68, 69, 70, 71, 72, 73, 74, 75, 77, 78, 79, 80, 83, 84, 85, 86, 87, 89, 90, 91, 94, 99, 102, 109, 110, 111, 112, 113, 114, 115, 116, 117, 118, 119, 121, 122, 123, 124, 125, 126, 127, 129, 130, 131, 142, 143, 145, 150, 154, 155, 157, 168, 170, 171
cancer cells, vii, x, 31, 32, 34, 35, 41, 46, 47, 51, 53, 55, 56, 62, 68, 69, 71, 72, 74, 75, 91, 109, 113, 115, 117, 118, 122, 123, 124, 125, 127
cancer death, viii, 37, 45, 60, 78, 110, 116
cancer therapy, x, 44, 47, 57, 58, 63, 109
carbohydrates, 95, 132, 137
carbon, 7, 8, 18, 37, 64, 139, 140
carboxyl, 50, 82, 102, 112, 149
carboxylic acid, 7, 158, 162
carcinogenesis, 49, 63, 73, 74, 80, 87, 123, 126, 150
carcinoma, viii, 32, 39, 46, 48, 70, 72, 80, 87, 88, 91, 123, 124, 125, 126, 150
cardiovascular disease, ix, 44, 46, 62, 71, 77, 79, 99, 124, 155
cell apoptosis, viii, xi, 32, 46, 79, 110
cell cycle, viii, 32, 40, 53, 54, 68, 80, 114, 115, 116, 118, 121, 124, 126
cell cycle arrest, viii, 32, 54, 80, 114, 116, 124, 126
cell death, 32, 40, 48, 53, 54, 55, 56, 58, 59, 68, 80, 101, 114, 115, 117, 118, 119, 120, 121, 125, 127, 156
cell line, ix, xi, 34, 35, 46, 52, 58, 59, 60, 66, 68, 69, 70, 75, 105, 112, 113, 114, 116, 122, 126, 129, 130, 131, 136, 142, 157
cell signaling, 95, 99, 125, 157
cervical cancer, 118, 125, 127, 157
challenges, 46, 73, 126
chemical, vii, 3, 5, 6, 31, 32, 34, 50, 51, 53, 56, 64, 66, 73, 74, 82, 84, 96, 100, 102, 103, 104, 111, 112, 113, 123, 126, 130, 131, 132, 138, 149, 156, 158, 160, 161
chemistry, v, vii, 1, 3, 6, 21, 22, 23, 24, 25, 26, 28, 33, 42, 50, 51, 57, 58, 59, 65, 72, 73, 74, 87, 105, 107, 113, 124, 125, 127, 144, 145, 146, 147, 160, 161, 168, 171, 172, 173, 174
chemotherapy, vii, x, 32, 35, 36, 40, 41, 47, 48, 55, 62, 63, 72, 73, 109, 110, 111, 122
chitosan, 59, 67, 68, 70, 72, 74, 83, 125, 143, 145, 146
chloroform, 3, 13, 28, 64, 132, 136, 137
cholesterol, 79, 98, 99, 154
chromatographic technique, 8, 11, 144
chromatography, 8, 11, 13, 18, 22, 23, 26, 27, 28, 29, 132, 133, 138, 145

Index

classification, 43, 48, 60, 104
co-crystals, xii, 148, 161, 162, 165, 168, 171
colo 205 cells, vi, xi, 129, 130, 131, 141, 157
colon, xi, 62, 66, 68, 69, 72, 73, 74, 75, 78, 87, 99, 110, 114, 120, 123, 124, 130, 136, 142, 143, 157
colon cancer, xi, 66, 68, 69, 72, 74, 75, 78, 114, 123, 124, 130, 142, 143, 157
colon carcinogenesis, 73, 74, 87
colorectal cancer, ix, xi, 59, 62, 63, 66, 67, 68, 70, 72, 74, 78, 114, 125, 129, 143
complications, 83, 94, 96, 101, 102, 104, 105
composition, 18, 21, 107, 160, 161
compounds, 5, 8, 11, 13, 18, 22, 28, 29, 33, 47, 49, 51, 73, 74, 81, 84, 87, 88, 89, 90, 94, 95, 100, 102, 105, 106, 107, 111, 113, 130, 133, 134, 135, 138, 144, 146, 149, 155, 158, 160
constituents, 23, 95, 99, 107, 145, 149, 150, 154
COOH, 2, 3, 18, 33, 140, 160
copolymer, 18, 65, 75, 81, 131
cosmetic, 103, 150, 157
crystal structure, 141, 159, 161
crystalline, xii, 3, 64, 78, 141, 148, 158, 159, 160, 161, 164, 166
crystalline solids, xii, 148, 158, 160, 161, 164
crystals, xii, 33, 148, 160, 161, 163, 164, 165, 166, 168, 171
curcumin, 28, 75, 81, 88
cure, 122, 150, 153
cyclooxygenase, 44, 81, 103, 124
cytochrome, 34, 40, 86, 115, 118, 120
cytokines, 85, 86, 97, 157
cytotoxic activity, xi, 129, 142
cytotoxicity, 35, 50, 52, 70, 117, 130, 157

D

death, vii, viii, ix, 31, 32, 37, 40, 43, 45, 47, 48, 53, 54, 55, 56, 57, 58, 59, 60, 61, 62, 68, 71, 78, 80, 95, 97, 99, 101, 110, 114, 115, 116, 117, 118, 119, 120, 121, 122, 125, 127, 156
defence, 37, 85, 96, 112
degradation, 41, 95, 99, 100, 116, 120
derivatives, ix, x, 3, 23, 33, 34, 43, 50, 51, 53, 57, 77, 79, 82, 84, 87, 93, 98, 100, 102, 104, 111, 112, 113, 115, 119, 123, 124, 125, 127, 149, 150, 151, 156
detection, 6, 114, 133, 135
diabetes mellitus (DM), x, 35, 42, 44, 78, 79, 83, 84, 85, 93, 94, 95, 96, 97, 99, 101, 102, 103, 104, 105, 106, 153, 155
dialysis, xi, 129, 136
differential scanning calorimetry (DSC), xi, 69, 70, 129, 135
diffraction, xi, 69, 129, 158, 159
disease, v, vii, viii, ix, x, 23, 27, 32, 34, 37, 38, 39, 43, 44, 46, 49, 57, 62, 63, 71, 77, 78, 79, 80, 84, 85, 89, 95, 98, 99, 102, 106, 107, 113, 123, 124, 150, 153, 155, 156, 157, 165
disorder, 79, 84, 156
dispersion, 74, 130, 131
distribution, 36, 37, 55, 63, 130, 131
DNA, 34, 51, 52, 60, 82, 90, 95, 99, 112, 115, 118, 125
DNA damage, 34, 52, 115, 118, 125
dosage, 36, 37, 63, 78
drug delivery, 65, 72, 74, 143
drug discovery, 22, 26, 95
drug release, xi, 70, 129, 136, 141
drug resistance, viii, 46, 47, 62
drugs, viii, ix, x, xi, 22, 23, 24, 25, 26, 27, 32, 34, 35, 36, 42, 44, 46, 47, 49, 50, 51, 57, 59, 60, 62, 63, 64, 65, 66, 68, 69, 70, 71, 72, 73, 74, 75, 79, 85, 87, 88, 89, 94, 95, 96, 105, 106, 109, 110, 111, 119, 125, 129, 130, 136, 141, 143, 145, 148, 157, 158, 161, 165, 166, 167, 168, 172

E

electron, 23, 69, 82, 132, 155, 162
elucidation, 7, 8, 23, 159

Index

emulsions, 65, 71, 135
encapsulation, 63, 65, 70, 71, 75, 96, 141
entrapment, xi, 70, 129, 145
environment, 35, 49, 51, 84, 95, 130, 136
enzymes, 33, 36, 42, 51, 64, 70, 73, 74, 80, 84, 85, 86, 88, 91, 95, 97, 112, 113, 157, 164
ester, 33, 34, 50, 51, 57, 65, 66, 82, 87, 112, 113, 115, 119, 125, 140, 150
ethanol, 3, 28, 33, 132, 134, 136, 137, 139
ethyl acetate, 3, 13, 132, 134, 136, 137, 138
evidence, ix, 46, 84, 100, 108, 114, 120
exposure, 49, 68, 79
extraction, 22, 25, 26, 33, 41, 42, 136, 138, 144, 145, 146
extracts, 3, 27, 29, 73, 87, 107, 121, 123, 132, 133, 139, 145, 148, 149, 151, 152, 172

F

fat, 84, 85, 102
flavonoids, 22, 26, 44, 83, 94, 107, 123, 132, 144, 145, 149, 155
flowers, 9, 29, 98
food, ix, 22, 23, 26, 27, 33, 34, 36, 37, 42, 43, 50, 51, 57, 59, 61, 63, 71, 73, 74, 75, 77, 78, 79, 81, 84, 85, 87, 88, 90, 98, 100, 101, 103, 104, 106, 107, 108, 111, 112, 113, 124, 126, 143, 144, 145, 146, 149, 150, 155, 158, 168, 169, 170, 171
food additive, 51, 79, 113, 158
formation, xii, 64, 66, 69, 82, 88, 95, 98, 107, 119, 122, 148, 156, 158, 159, 161, 162, 163, 166
formula, 3, 33, 64, 84, 98, 99, 139
fourier transform infrared spectroscopy (FTIR), 2, 6, 24, 27, 69, 75, 123, 133, 135, 139, 140
free radicals, 26, 79, 83, 86, 94, 101, 112, 130, 139, 149, 155, 165
fruit, ix, xi, 23, 27, 28, 49, 59, 67, 70, 72, 74, 77, 78, 81, 84, 98, 100, 106, 107, 125, 130, 131, 132, 133, 134, 136, 137, 138, 139, 143, 144, 146, 147, 149, 151, 152, 153, 154, 155, 168

G

gallic acid-loaded sodium alginate, xi, 69, 72, 129
Gallo tannins, ix, 77, 97, 99
gel, 11, 12, 13, 14, 15, 16, 17, 137
gene expression, 52, 75, 85, 90, 91, 119, 120, 121
genes, 48, 54, 58, 85, 97, 114, 115, 116, 117, 122
glucose, 35, 83, 94, 95, 96, 97, 100, 102, 103, 104, 105, 108, 111, 130
GLUT4, 97, 100, 104
glutathione, 34, 59, 80, 97, 117, 120, 125, 127, 155, 157
glycerol, 3, 33, 64, 111, 139
glycogen, 44, 83, 94
gold nanoparticles, ix, 40, 62, 63
growth, vii, 31, 40, 46, 47, 51, 52, 53, 54, 55, 56, 58, 59, 65, 66, 69, 80, 86, 88, 107, 113, 116, 118, 120, 121, 126, 142, 149

H

health, vii, ix, x, xi, 22, 32, 57, 58, 62, 77, 80, 83, 84, 93, 94, 102, 104, 107, 114, 147, 149, 150, 157, 168
hepatitis, ix, 44, 62, 71, 82, 89, 104, 110, 124
hepatocellular carcinoma, 70, 80, 87, 91, 125, 126
hepatocytes, 89, 97, 104
high performance liquid chromatography (HPLC), 1, 2, 6, 10, 18, 19, 21, 22, 23, 24, 25, 26, 27, 28, 29, 30, 134, 137, 138, 139, 144
high performance thin layer chromatography (HPTLC), 1, 2, 13, 14, 21, 22, 23, 24, 25, 26, 27, 28, 29, 42, 134, 137
HIV, 48, 82, 89, 155, 158
homeostasis, 79, 83, 94, 96, 100

hormone, vii, 32, 40
hormone therapy, vii, 32, 40
host, 79, 95, 157
human, viii, ix, xi, 32, 35, 36, 40, 41, 42, 44, 46, 55, 59, 60, 68, 69, 70, 75, 81, 82, 84, 87, 88, 89, 91, 99, 100, 101, 103, 104, 105, 107, 110, 113, 114, 117, 118, 120, 122, 123, 124, 125, 126, 127, 129, 130, 142, 143, 150, 156, 167
hydrazine, 73, 74, 87
hydrogen, xii, 50, 97, 112, 117, 140, 148, 155, 159, 160, 161, 162, 163, 166
hydrogen bonds, 50, 159, 160
hydrogen peroxide, 97, 117, 155
hydrolysable tannins, ix, 28, 33, 77, 150, 151
hydrolysis, xi, 36, 99, 147, 151
hydroxyl, 50, 82, 111, 112, 139, 140, 155, 158, 159
hyperglycemia, 79, 83, 94, 95, 97, 99, 104

I

identification, 21, 23, 134
in vitro, ix, x, xi, 34, 41, 42, 51, 62, 68, 70, 71, 75, 77, 79, 87, 105, 108, 122, 129, 130, 131, 142, 143, 146, 156, 164, 167
in vivo, 34, 35, 68, 75, 79, 80, 87, 105, 106, 142, 146, 156
incidence, 46, 47, 48, 57, 72, 78, 119, 123
individuals, viii, 35, 45, 104
inducer, 49, 63, 99
induction, viii, ix, 32, 36, 46, 56, 58, 86, 105
industry, 49, 50, 64, 71, 78, 112, 150
infarction, 86, 88, 91
infection, 34, 37, 89
inflammation, ix, x, 35, 62, 71, 84, 85, 86, 90, 91, 93, 95, 100, 103, 104, 106, 108, 120, 153, 154, 165
inflammatory disease, 79, 85, 99
infrared spectroscopy (IR), xi, 2, 6, 7, 21, 129, 133, 135, 137, 138, 139

inhibition, viii, ix, 32, 46, 52, 54, 55, 59, 75, 89, 90, 91, 101, 115, 118, 122, 125, 164
inhibitor, ix, 46, 51, 52, 54, 55, 116, 122
injury, 79, 95, 96, 99
insulin, 27, 44, 79, 83, 90, 94, 95, 97, 99, 100, 103, 104, 105, 106, 108, 124
insulin resistance, 79, 95, 97, 99, 106
insulin sensitivity, 95, 97, 99, 100, 108
ions, 49, 63, 105, 156
iron, ix, 62, 63, 158
ischemia, 78, 86, 143
issues, xii, 83, 94, 130, 148, 160, 163, 165

L

layered double hydroxides, ix, 62, 63
lead, viii, xi, 32, 41, 46, 50, 99, 106, 110, 111, 166
leukemia, 87, 91, 125
light, 5, 33, 96, 125, 130
lipid peroxidation, 41, 80, 83, 88, 91, 94, 101, 105, 117
lipids, 95, 96, 99, 102, 112
liposomes, 64, 65, 71
liquid chromatography, 8, 22, 23, 28, 29
liver, ix, 36, 38, 61, 71, 78, 80, 83, 84, 101, 102, 119, 153
liver cancer, ix, 61, 71, 119
lung cancer, viii, ix, 45, 46, 47, 48, 52, 54, 55, 56, 57, 58, 59, 60, 62, 66, 72, 75, 78, 86, 91, 110, 116, 117, 126, 127, 157
lymphoma, 34, 40, 53, 59, 114

M

macromolecules, 56, 65, 75
macrophages, 84, 85, 90, 165
majority, ix, 40, 47, 62, 71
malignancies, vii, viii, xi, 31, 40, 46, 48, 51, 52, 56, 109, 111, 113, 116, 122
management, 43, 44, 46, 56, 84, 87, 94, 105, 114, 116
mass, 8, 9, 18, 21, 23, 29, 37, 66, 85, 110
mass spectrometry (MS), 2, 8, 9, 10, 18, 21, 22, 23, 24, 28, 29, 43, 44, 145, 169

materials, xii, 3, 111, 148, 158, 160, 161, 164
matrix, 21, 36, 40, 69, 80, 125
matrix metalloproteinase (MMP), 40, 55, 56, 80, 117, 120, 121, 125, 126
matter, 74, 108, 156
MCF-7 cells, viii, 32, 40, 88, 115
measurement, 41, 99, 134
medical, 34, 103, 107
medication, 36, 56, 64, 65, 94, 110, 117, 130
medicine, viii, ix, 26, 28, 32, 35, 41, 47, 53, 57, 58, 59, 62, 63, 65, 71, 83, 87, 91, 94, 104, 106, 107, 111, 121, 126, 148, 149, 170, 171
melanoma, vii, 32, 110, 120, 125
mellitus, x, 42, 44, 79, 83, 93, 94, 95, 96, 97, 99, 102, 104, 105, 155
melting, 49, 64, 134, 140, 141, 158
membranes, 82, 83, 86
men, ix, 37, 47, 48, 61, 62, 71, 78, 119
metabolic disorder, 79, 84, 101
metabolic syndrome, 79, 97, 99, 104
metabolism, 42, 70, 79, 88, 95, 99, 102
metabolite, 29, 35, 36, 64, 78, 84, 87, 97, 98, 102, 108, 123, 132, 144, 150
metal ions, 49, 63, 105, 156
metalloproteinase, 40, 117, 125
metastasis, viii, 32, 40, 48, 74, 80, 121, 122, 157
methanol, 3, 6, 10, 16, 18, 29, 33, 132, 134, 137, 138, 139, 163
mice, 97, 102, 104, 108, 125, 126
microspheres, 69, 72, 75
migration, viii, xi, 32, 40, 46, 71, 110, 117, 121, 122, 125, 157
mitochondria, 69, 90, 107, 115, 119, 121
mitochondrial pathway, 32, 118, 126
MMP-9, 40, 117, 126
models, x, 35, 36, 77, 80, 141
modifications, 46, 51, 113, 115, 117, 159
molecular weight, vii, xi, 1, 33, 64, 78, 99, 111, 147
molecules, 8, 18, 53, 56, 85, 95, 112, 149, 158, 159, 160, 161, 167
morbidity, x, 46, 47, 93
mortality, vii, x, 31, 34, 37, 40, 43, 46, 47, 54, 55, 57, 72, 78, 93, 109, 113, 122, 123
mutant, 51, 52, 58, 116
mutations, 47, 48, 52, 54
myocardial infarction, 86, 88, 91

N

nanocarriers, ix, 62, 63, 66, 75
nanoformulations, v, ix, 61, 62, 63, 64, 65, 66, 70, 71, 100, 141, 142, 143
nanomaterials, 72, 111, 135
nanoparticles, vi, ix, xi, 40, 59, 62, 63, 64, 65, 67, 68, 69, 70, 71, 72, 74, 75, 90, 96, 125, 129, 130, 131, 135, 136, 140, 141, 142, 143, 145, 146
nanotechnology, ix, 62, 73, 111
necrosis, 68, 85, 90, 103, 114
neurodegenerative diseases, 84, 150, 156
neutral, 12, 100, 130, 161
nodes, 37, 38, 172
non-small-cell lung cancer (NSCLC), viii, ix, 45, 46, 48, 52, 53, 54, 56, 58, 116
nuclear magnetic resonance (NMR), 1, 2, 7, 8, 21, 23, 25, 30, 102, 133, 138, 139
nucleus, 71, 117, 122
nutraceutical, 67, 74, 102
nutrition, 37, 75, 84, 90, 107

O

obesity, viii, x, 32, 77, 78, 84, 95, 153
oil, 49, 126, 152
optimization, 78, 140, 145
organs, ix, 39, 47, 61, 71, 78, 79, 95, 96
oxidation, 23, 33, 34, 50, 56, 79, 95, 96, 99, 102, 105, 112, 149
oxidative damage, 36, 81, 95, 99, 101, 121, 130
oxidative stress, ix, x, 55, 75, 77, 78, 79, 83, 93, 94, 95, 96, 97, 99, 100, 101, 102, 103, 104, 106, 108, 112, 117, 120, 143, 150, 155, 156, 157, 166

oxygen, 7, 34, 37, 40, 50, 81, 85, 86, 95, 96, 112, 117, 125, 127, 149, 155

P

p53, 40, 54, 55, 57, 114, 117, 143
pathogens, 89, 103, 157, 167
pathways, viii, x, xi, 3, 4, 24, 32, 40, 41, 46, 50, 51, 52, 53, 55, 58, 60, 63, 64, 79, 80, 87, 90, 91, 95, 96, 97, 100, 101, 102, 104, 105, 107, 108, 109, 110, 113, 114, 117, 118, 119, 120, 121, 122, 124, 126, 157, 162
peroxidation, 41, 80, 83, 88, 91, 94, 95, 99, 101, 105, 117
peroxide, 59, 97, 117, 119, 126, 155
pH, xi, 19, 20, 69, 75, 129, 130, 136, 141, 143, 165
pharmaceutical, ix, 3, 22, 26, 34, 50, 62, 63, 64, 65, 71, 78, 103, 106, 111, 112, 148, 150, 160, 161, 163
pharmacognosy, vi, 26, 27, 131, 147, 148, 170, 171
pharmacological research, xi, 148, 163
pharmacology, 60, 73, 75, 124, 130
phenolic compound, vii, viii, ix, 1, 3, 22, 28, 31, 32, 33, 70, 77, 78, 84, 88, 95, 96, 98, 107, 110, 113, 149
phosphate, 19, 83, 94, 136
phosphorylation, 40, 58, 113, 122
physicochemical properties, 70, 113, 130
PI3K, 41, 53, 75, 97, 107, 111, 121, 122
plants, vii, viii, ix, x, xi, 1, 3, 21, 24, 31, 32, 33, 34, 42, 47, 49, 50, 53, 56, 57, 58, 59, 63, 64, 73, 77, 78, 81, 84, 87, 88, 93, 97, 98, 99, 101, 102, 104, 105, 107, 111, 112, 117, 131, 132, 133, 135, 143, 145, 147, 148, 149, 150, 151, 152, 153, 154, 155, 169, 172
polar, 3, 6, 11, 13, 18, 33, 82
polarity, 13, 100, 146
polymers, ix, 18, 62, 63, 65, 66, 68, 70, 74, 75, 130, 172
polyphenols, viii, 21, 23, 29, 42, 46, 49, 57, 58, 59, 64, 78, 79, 81, 84, 87, 88, 97, 98, 101, 104, 106, 111, 113, 116, 117, 124, 126, 132, 145, 149, 155
polysaccharide, 95, 96, 105
population, 37, 41, 57, 115
preparation, iv, 13, 63, 113, 131, 160, 166
prevention, x, 41, 53, 83, 84, 87, 90, 94, 109, 110, 150
probe, 67, 70, 135
prognosis, 48, 119, 121
proliferation, viii, xi, 32, 41, 46, 47, 52, 56, 68, 80, 85, 86, 87, 110, 113, 114, 115, 121, 122, 127, 142, 143
prostate cancer, 58, 60, 86, 115, 123
protection, 70, 83, 94
proteins, 56, 80, 91, 95, 96, 97, 99, 112, 114, 115, 118, 120, 121, 156
protons, 3, 7, 21, 139, 162
public health, 62, 72, 83, 94
purity, 13, 134, 140

Q

quantification, 5, 8, 13, 18, 21, 23, 25, 27, 29, 73, 87, 107
quercetin, 29, 36, 59, 67, 68, 72, 74, 84, 125, 130, 143

R

radiation, vii, 32, 40, 47, 56, 103, 110, 122
radicals, 23, 26, 79, 83, 86, 94, 101, 104, 112, 120, 130, 139, 149, 155, 166
radiotherapy, 32, 47, 62
reactive oxygen, 40, 85, 95, 112, 117, 125, 127, 155
receptor, 47, 48, 53, 54, 58, 59, 97, 105, 116, 119
red wine, 22, 50, 63, 78, 101
regression, 120, 138, 139
researchers, 33, 34, 50, 64, 121, 160, 161
resistance, viii, ix, x, 46, 47, 54, 56, 58, 59, 62, 79, 80, 95, 97, 99, 106, 109, 117, 158
resolution, 21, 133, 144
response, 48, 49, 52, 83, 85, 90, 97, 108, 115, 142, 156, 157

risk, 47, 48, 85, 86, 119
root, 25, 63, 71, 144, 152, 153, 154, 155

S

salts, 69, 99, 160, 162, 165
scavengers, 81, 100, 155
science, 58, 74, 75, 107, 108, 143, 145, 148
secretion, 79, 83, 94
seed, 26, 84, 89, 105, 142, 152, 154
selectivity, 18, 21, 71
sensitivity, 6, 13, 18, 21, 46, 52, 58, 95, 97, 99, 100, 108, 117
serum, 35, 81, 86, 99, 101, 136
shape, 38, 56, 64, 74, 121
side effects, vii, viii, xi, 32, 46, 53, 62, 123, 148, 163
signaling pathway, 40, 52, 53, 60, 79, 87, 97, 101, 105, 107, 108, 117, 121, 122
signalling, 40, 58, 79, 87, 95, 96, 99, 100, 101, 116, 120, 121
signals, 75, 85, 123
signs, 34, 38, 83, 94, 121
silica, 11, 13, 14, 15, 16, 17, 18, 133, 137
skin, 38, 39, 51, 110, 125, 126, 153, 154, 157
small cell lung cancer (SCLC), viii, ix, 45, 46, 48, 52, 54, 55, 58, 59, 60, 72, 79, 116, 126
smoking, 46, 48, 52, 110
sodium, xi, 69, 72, 75, 129, 130, 131, 135, 140, 141
solubility, xii, 3, 13, 23, 49, 51, 63, 65, 96, 112, 113, 130, 134, 139, 148, 158, 161, 163, 164, 165, 166
solution, 20, 34, 50, 67, 112, 132, 133, 134, 135, 136, 137, 138, 139, 166
solvents, 3, 6, 11, 13, 33, 131, 132, 134, 136, 137, 139, 145
species, xi, 10, 23, 28, 40, 49, 63, 82, 85, 97, 98, 112, 117, 125, 127, 130, 147, 151, 153, 155, 165
spectroscopy, xi, 1, 2, 5, 6, 7, 8, 21, 22, 23, 123, 129, 133, 135, 137, 138, 139, 144, 173

stability, xii, 70, 71, 75, 87, 90, 96, 130, 148, 158, 161, 163, 166
state, viii, 43, 45, 46, 47, 57, 95, 99, 108, 110, 116, 158, 160
statistics, 43, 58, 72, 87
stomach, ix, 61, 71, 78, 110, 153, 154, 157
stress, ix, x, 55, 75, 77, 78, 79, 83, 93, 94, 95, 96, 97, 99, 100, 101, 102, 103, 104, 106, 108, 112, 117, 120, 143, 150, 155, 156, 157, 166
structure, xii, 5, 23, 43, 51, 64, 82, 100, 101, 105, 112, 113, 141, 148, 149, 151, 158, 159, 160, 167
substrate, 34, 50, 112, 122; 149
sulfuric acid, 34, 50, 112
sun, 19, 28, 29, 145, 146
supplement, x, 50, 78, 87, 95, 103, 104, 111, 157
suppression, 54, 60, 69, 116, 118, 120, 122, 143
survival, 52, 54, 80, 116, 118, 119, 121, 142
symptoms, 36, 38, 83, 85, 94
syndrome, 79, 97, 99, 104
synthesis, 34, 50, 51, 64, 65, 66, 71, 90, 104, 106, 112, 120, 161

T

tannins, ix, 3, 25, 28, 33, 49, 50, 77, 97, 98, 99, 111, 112, 132, 137, 150, 151
target, x, 54, 62, 79, 85, 95, 109, 110, 161
techniques, vii, 1, 3, 8, 11, 13, 21, 64, 65, 74, 119, 145
technology, 42, 74, 138
temperature, 9, 10, 64, 133, 135, 139, 141, 165
therapeutic effect, viii, xii, 32, 113, 148, 164
therapy, vii, viii, x, 32, 40, 44, 46, 47, 52, 55, 56, 57, 58, 60, 63, 77, 78, 81, 89, 99, 100, 101, 102, 104, 109, 110, 114, 115, 116, 118, 119, 122
thermal stability, xii, 148, 163

thin layer chromatography (TLC), 2, 11, 12, 13, 22, 23, 25, 26, 27, 28, 29, 133, 134, 138, 145
thyroid cancer, ix, 61, 71
tissue, ix, 37, 38, 39, 47, 61, 63, 71, 78, 79, 83, 85, 94, 95, 96, 100, 102, 110, 120, 165
TNF, 85, 90, 100, 103, 117
tobacco, 47, 48, 51, 52, 57
toluene, 13, 134, 138
tonic, 33, 153, 154
toxicity, viii, xi, 35, 36, 46, 47, 50, 56, 72, 75, 79, 100, 125, 148, 163
toxicology, 60, 73, 74, 124, 126
transcription, 52, 85, 119, 122
translocation, 97, 100, 117, 120
transport, 37, 64, 65, 82, 96, 105, 132, 141
treatment, viii, ix, x, 32, 34, 40, 41, 43, 46, 47, 52, 53, 55, 57, 58, 62, 63, 66, 67, 71, 72, 74, 78, 83, 84, 85, 86, 89, 90, 95, 96, 99, 101, 103, 104, 109, 110, 113, 115, 116, 117, 118, 119, 120, 121, 122, 126, 132, 142, 143, 153, 157
tuberculosis, 73, 153, 165
tumor, vi, vii, viii, x, xi, 31, 32, 37, 38, 41, 44, 46, 47, 48, 49, 52, 53, 54, 56, 58, 60, 68, 78, 80, 85, 86, 90, 91, 99, 103, 107, 109, 110, 111, 113, 114, 116, 117, 119, 120, 121, 122, 123, 125, 126, 150

turnover, 52, 58, 116
tyrosine, ix, 46, 48, 52, 54, 58, 59, 116, 119

V

validation, 22, 25, 26, 27, 28, 29, 138
valuation, 106, 143, 146
vegetables, 101, 107, 149
VEGF, 113, 122, 126
vessels, 37, 38, 122

W

water, 3, 6, 9, 19, 33, 34, 49, 50, 64, 65, 68, 100, 112, 130, 135, 137, 141, 161, 165, 166
women, vii, ix, 31, 37, 40, 47, 48, 61, 62, 71, 78
wood, xi, 49, 63, 147
World Health Organization (WHO), 37, 48, 60, 79, 114
worldwide, vii, viii, ix, 31, 40, 45, 61, 62, 71, 72, 78, 83, 94, 116, 123

X

X-ray diffraction (XRD), xi, 69, 70, 75, 129, 158, 159